PPh 2
Progress in Physics

Edited by A. Jaffe and D. Ruelle

VORTICES AND MONOPOLES

Structure of Static Gauge Theories

Arthur Jaffe
Clifford Taubes

BIRKHÄUSER
BOSTON • BASEL • STUTTGART

Authors

Arthur Jaffe
Harvard University
Cambridge, MA 02138

Clifford Taubes
Department of Physics
Harvard University
Cambridge, MA 02138

Library of Congress Cataloging in Publication Data

Jaffe, Arthur, 1937-
Vortices and monopoles.
(Progress in physics ; 2)
Bibliography: p.
Includes index.
1. Gauge fields (Physics) 2. Topology. 3. Numerical analysis.
I. Taubes, Clifford, 1954- joint author. II. Title.
QC793.3.F5J33 530.1'4 80-20080
ISBN 3-7643-3025-2

CIP—Kurztitelaufnahme der Deutschen Bibliothek

Jaffe, Arthur
Vortices and monopoles : structure of static gauge theories/
Arthur Jaffe ; Clifford Taubes. — Boston, Basel, Stuttgart :
Birkhäuser, 1980.
(Progress in physics ; 2)
ISBN 3-7643-3025-2

NE: Taubes, Clifford:

ISBN: 3-7643-3025-2

Printed in USA

TABLE OF CONTENTS

TABLE OF CONTENTS (continued)

PREFACE

Gauge theories arise in the description of electromagnetic phenomena, including superconductivity. They also have become central to elementary particle physics, as the starting point for a quantum theory of the weak, strong and electromagnetic interactions. In the past few years classical gauge theories have been widely studied by topologists and geometers interested in their underlying structure. One outcome of the work of physicists and mathematicians has been the discovery of all "instantons," as well as a partial understanding of their role in quantum theory.

We present here another point of view. A combination of analytic and topological methods can give useful insights into classical gauge theories for which purely geometric methods have yet to be successful. In particular we use these methods to analyze general features of the Ginzburg-Landau equations for vortices and the Yang-Mills-Higgs equations for monopoles, for which solutions in closed form have not been found.

In Chapters I-II we give some general background for gauge theories. Next we treat specific problems of finding and characterizing solutions to the vortex and monopole equations. The methods of Chapters III-V are general and surely have wider applicability to equations not studied here. In Chapter VI we collect together some analytic tools which can be useful in a variety of problems.

We thank D. Groisser for many suggestions and for his assistance in proof reading. We are especially grateful to Renate D'Arcangelo for artistic typing under pressure. Her heroic effort made this publication possible.

We have attempted to provide a timely introduction to the mathematical analysis of classical gauge theories. We hope both theoretical physicists and mathematicians will find this work useful.

A. Jaffe, C. Taubes
Cambridge, Massachusetts
August, 1980

I. GAUGE THEORIES

This introductory chapter serves as a general orientation for the mathematical development which begins in Chapter II. We present the basic equations studied throughout and give a brief account of how they are used by physicists.

I.1. YANG-MILLS-HIGGS THEORY

The dynamical variables for a Yang-Mills-Higgs theory with d space dimensions are a gauge potential (connection)

$$A = A_j(x)dx^j$$

and a scalar (Higgs) field[1]

$$\Phi = \Phi(x)$$

In one time plus d space dimensions the variables are similar. The components $A_j(x)$ take their values in the Lie algebra $\mathfrak{g}$ of a matrix Lie group G called the gauge group. The field Φ takes its values in a vector space $\mathbb{L}$ (the internal symmetry space) on which G acts as a transformation group.

The Yang-Mills potential defines a field (the curvature)

$$F = dA + A \wedge A = \frac{1}{2} F_{ij}(x) dx^i \wedge dx^j \tag{1.1}$$

---

[1] What physicists refer to as a field is what mathematicians refer to as a section of a vector bundle over $\mathbb{R}^d$. The orthogonal group $O(d)$ acts on sections of such a bundle. If the group acts trivially, the field is called *scalar*. Otherwise it is called *vector* or *tensor*. The connection A, for example, is called a vector field. This is not be confused with what mathematicians call a vector field.

with components

$$F_{ij}(x) = \partial_i A_j(x) - \partial_j A_i(x) + [A_i(x), A_j(x)] \quad . \tag{1.2}$$

Here as in the following we use the summation convention.

The Higgs field couples to the connection through the covariant derivative. This is called "minimal coupling." Let ρ be a linear representation of G on $\mathbb{L}$. Then ρ induces a representation of the Lie algebra $\mathcal{g}$ on $\mathbb{L}$, and the covariant derivative is defined by

$$(\nabla_A)_j(\Phi) = \nabla_j \Phi + \rho(A_j)(\Phi) \tag{1.3a}$$

and

$$D_A \Phi = (\nabla_A)_j(\Phi) dx^j \quad . \tag{1.3b}$$

In Chapter III, we consider the abelian Higgs model on $\mathbb{R}^2$, with the gauge group $G = U(1)$. Its Lie algebra is $\mathcal{g} = -i\mathbb{R}$ and we choose the internal symmetry space $\mathbb{L} = \mathbb{C}$, the complex numbers. The group $U(1)$ acts on $\mathbb{C}$ as multiplication by the group of complex numbers of unit modulus.

In the physics literature, when $G = U(1)$, the connection is written $-iA$ with A real. Likewise the curvature is written $-iF_A = -idA$. We adhere to this convention in Chapter III. Thus for Φ a $\mathbb{C}$-valued function on $\mathbb{R}^2$

$$(\nabla_A)_j(\Phi) = \nabla_j \Phi - iA_j \Phi \quad , \qquad A_j \in \mathbb{R} \quad . \tag{1.4}$$

In Chapter IV we consider the non-abelian Higgs model on $\mathbb{R}^3$. In this case G is a compact, connected Lie group, and we choose the vector space $\mathbb{L}$ to be $\mathcal{g}$. The action of G on $\mathcal{g}$ is the usual adjoint action. Thus the covariant derivative is

$$(\nabla_A)_j(\Phi) = \nabla_j \Phi + [A_j, \Phi] \quad . \tag{1.5}$$

We use the symbol $(\ ,\ )$ to denote a vector space inner product, whether the vector space is $\mathcal{g}$ or $\mathbb{L}$. No confusion should arise. Also $|a| \equiv (a,a)^{1/2}$. For example, if $\mathbb{L} = \mathbb{C}$ then

$$(a,b) = a\bar{b} \quad , \qquad a,b \in \mathbb{C} \ .$$

This is the inner product used in Chapter III.

In Chapter IV, where $\mathbb{L} = su(2)$, the inner product we use is

$$(a,b) = 2\mathrm{tr}(a^{\dagger}b) \equiv \mathrm{Tr}(a^{\dagger}b) , \qquad a,b \in su(2) \ .$$

Here $su(2)$ has been identified with the vector space of 2×2 anti-hermitian, traceless matrices. The symbol "$\dagger$" denotes the hermitian conjugate, "tr" denotes the usual matrix trace and "Tr" is the normalized trace. For $\mathbb{L} = \mathcal{g}$, the Lie algebra of a compact, simple Lie group G, the inner product $(\ ,\ )$ also is a suitably normalized trace, denoted Tr.

The inner product extends to $\mathbb{L}$-valued or $\mathcal{g}$-valued p-forms. Let η and ω be vector valued p-forms,

$$\eta = \eta_{i_1 \ldots i_p} \, dx^{i_1} \wedge \ldots \wedge dx^{i_p} ,$$

$$\omega = \omega_{i_1 \ldots i_p} \, dx^{i_1} \wedge \ldots \wedge dx^{i_p} .$$

Then

$$(\omega,\eta) \equiv \frac{1}{p!} \left(\omega_{i_1 \ldots i_p}, \eta_{i_1 \ldots i_p} \right) \quad . \tag{1.6a}$$

Furthermore, if ω is a vector-valued zero-form and η is a vector valued p-form, we let (ω,η) denote the scalar valued p-form

$$(\omega,\eta) \equiv \left(\omega,\eta_{i_1 \ldots i_p} \right) dx^{i_1} \wedge \ldots \wedge dx^{i_p} \quad . \tag{1.6b}$$

The Hodge operator $*$ on $\mathbb{R}^d$ maps p-forms to (d-p)-forms. With η a p-form as before,

$$*\eta = \frac{1}{p!} \varepsilon^{j_1 \dots j_p i_1 \dots i_{d-p}} \eta_{j_1 \dots j_p} dx^{i_1} \wedge \dots \wedge dx^{i_{d-p}} . \tag{1.7}$$

Here $\varepsilon^{i_1 \dots i_d}$ is the totally antisymmetric tensor and $\varepsilon^{1 \dots d} = 1$. Since we work on flat Euclidean space, raising or lowering indices has no significance. Note also

$$(\omega, \eta) = (*\omega, *\eta) ,$$

i.e., $*$ is an isometry.

With these preliminary definitions, the Euclidean Yang-Mills-Higgs action is

$$\mathscr{A}(A, \Phi) = \frac{1}{2} \int_{\mathbb{R}^d} \left\{ (F_A, F_A) + (D_A \Phi, D_A \Phi) + \frac{\lambda}{4} *(|\Phi|^2 - 1)^2 \right\} , \tag{1.8}$$

where $\lambda \geq 0$ is a constant. The term $(\lambda/8)(|\Phi|^2 - 1)^2$ is the Higgs self-interaction.

On Minkowski space, the action density is

$$a_M = \frac{1}{2} \left\{ \frac{1}{2} (F_{ij}, F_{ij}) - (F_{0i}, F_{0i}) + ((\nabla_A)_i \Phi, (\nabla_A)_i \Phi) - ((\nabla_A)_0 \Phi, (\nabla_A)_0 \Phi) + \frac{\lambda}{4} *(|\Phi|^2 - 1)^2 \right\} . \tag{1.9a}$$

Here the spatial indices are $i, j \in \{1, \dots, d\}$, and 0 denotes the time component. The action is

$$\mathscr{A}_M = \int_{\mathbb{R}^{d+1}} a_M . \tag{1.9b}$$

In case that the fields (A,Φ) are independent of time, and in addition that $A_0 = 0$, we say the field configuration (A,Φ) is *static*. In this case the integral (1.9b) does not exist. However, the time zero integral of a_M over the hyperplane $\mathbb{R}^d$, namely

$$\mathscr{A} = \int_{\substack{\mathbb{R}^d \\ x_0=0}} a_M \quad , \tag{1.9c}$$

is just (1.8). The expression (1.9c) is called the *energy* of the static configuration. In general, the static energy equals the Euclidean action in one lower dimension.

As the title of this book suggests, we are concerned almost exclusively with the static problem, namely the action (1.8) and its variational equations. Henceforth $\mathscr{A}$ means (1.8).

The relative normalization of $(\ ,\ )$ on $\mathbb{L}$ and on $\mathcal{g}$ is not important in (1.8). This relative normalization can be changed by the scale transformation

$$\begin{aligned} A_j(x) &\to A_j'(x) = cA_j(cx) \\ \Phi(x) &\to \Phi'(x) = \Phi(cx) \\ x &\to x' = xc^{-1} \\ \lambda &\to \lambda' = \lambda c^d \quad . \end{aligned} \tag{1.10}$$

The Euclidean action becomes

$$\mathscr{A}(A',\Phi';\lambda') \quad \frac{1}{2} \int_{\mathbb{R}^d} \left\{ c^{4-d}(F_A,F_A) + c^{2-d}(D_A\Phi,D_A\Phi) + \frac{\lambda}{4} *(|\Phi|^2 - 1)^2 \right\} \tag{1.11}$$

Here we emphasize the dependence of the action on the constant λ by writing $\mathscr{A} = \mathscr{A}(A,\Phi;\lambda)$.

Given a representation of a compact Lie group on a vector space, it is always possible to choose an inner product (,) with respect to which the representation is unitary. We therefore restrict attention to unitary representations.

We now explain why the action $\mathscr{A}$ is invariant under smooth gauge transformations. Except in Chapter V, where we consider the Yang-Mills-Higgs equations on an arbitrary Riemannian manifold, it is sufficient to define a gauge transformation as a smooth map $g: \mathbb{R}^d \to G$. The gauge transformation g acts on the fields in the usual way:

$$A \to A_g = gAg^{-1} + gdg^{-1} \quad ,$$

$$\Phi \to \Phi_g = \rho(g)(\Phi) \quad ,$$

$$F_A \to F_{A_g} = gF_A g^{-1} \quad ,$$

and

$$D_A\Phi \to D_{A_g}\Phi_g = \rho(g)(D_A\Phi) \quad . \tag{1.12}$$

From (1.12), it follows that $\mathscr{A}$ is gauge invariant.

In order to write the variational equations for $\mathscr{A}$ in compact form, we need the general covariant exterior derivative D_A. Suppose ω is an $\mathbb{L}$-valued p-form. Then D_A is defined by

$$D_A\omega \equiv d\omega + \rho(A) \wedge \omega \quad . \tag{1.13}$$

This derivative (1.13) is covariant in the sense that

$$D_{A_g}(\rho(g)\omega) = \rho(g)(D_A\omega) \quad . \tag{1.14}$$

Similarly if ω is a $\mathcal{g}$-valued p-form,

$$D_A\omega \equiv d\omega + A \wedge \omega - (-1)^p \omega \wedge A \quad . \tag{1.15}$$

Then (1.15) is covariant in the sense that

$$D_{A_g}(g\omega g^{-1}) = g(D_A\omega)g^{-1} \quad . \tag{1.16}$$

Furthermore in the special case that ρ is the adjoint representation acting on $\mathbb{L} = \mathfrak{g}$, then (1.13) and (1.15) agree.

We also define the operator $(\nabla_A)_j$ on p-forms by

$$(\nabla_{A_j})(\omega) = (\nabla_A)_j(\omega_{i_1\ldots i_p})dx^{i_1}\wedge\ldots\wedge dx^{i_p} \quad ,$$

with $(\nabla_A)_j$ given in (1.3a). Then

$$\nabla_A^2 \equiv (\nabla_A)_j(\nabla_A)_j \quad .$$

We now state the variational equations for (1.8). They take the form

$$D_A{*F} = {*J} \quad , \tag{1.17a}$$

$$\nabla_A^2\Phi = \frac{\lambda}{2}\Phi(|\Phi|^2 - 1) \quad . \tag{1.17b}$$

For the case $G = U(1)$, $\mathbb{L} = \mathbb{C}$, and $\rho(A)$ is multiplication by $-iA$ as in (1.4), the current J is given by

$$J = (i\Phi, D_A\Phi) \quad . \tag{1.18a}$$

For G arbitrary, $\mathbb{L} = \mathfrak{g}$, and ρ the adjoint representation, the current is

$$J = -[\Phi, D_A\Phi] \quad . \tag{1.18b}$$

For a general group G we introduce an orthonormal basis $\{h^a\}$ for the Lie algebra $\mathfrak{g}$. Then equations (1.18) are special cases of the general expression

$$J = -(\rho(h^a)\Phi, D_A\Phi)h^a \quad . \tag{1.17c}$$

These equations (1.17) are the subject of our study.

I.2. SOLITONS

In the next three sections we describe some qualitative properties of the solutions to the equations (1.17). Finite action solutions are called *solitons*. As explained in §1, these are time-independent finite energy solutions to the variational equations for the action density (1.9a) on (d+1)-dimensional Minkowski space.

The most familiar examples of gauge theory solitons are the "instantons." These are finite action solutions to the $d = 4$ Euclidean "pure" Yang-Mills equations $(\lambda = 0,\ \Phi \equiv 0)$. The instantons have the property that their curvatures are self-dual. In terms of the Hodge duality operator $*$ defined in (1.7), the self-duality condition means that

$$F_A = dA + A \wedge A$$

satisfies

$$*F_A = F_A \quad \text{or} \quad *F_A = -F_A \quad . \tag{2.1}$$

This relation ensures that F_A satisfies the variational equations (1.17), because every curvature F_A automatically satisfies the Bianchi identity

$$D_A F_A = 0 \quad . \tag{2.2}$$

This follows from $d^2 = 0$ and the definitions of D_A and F_A. Clearly, (2.1-2) imply

$$D_A * F_A = 0 \quad . \tag{2.3}$$

This is equation (1.17) in the case $\Phi = 0$. To summarize, every curvature F_A that satisfies the self-duality condition (2.1) is a critical point of the $d = 4$ Yang-Mills action

$$\mathscr{A}_{\text{Yang-Mills}} = \frac{1}{2} \int_{\mathbb{R}^d} (F_A, F_A) \ .$$

All finite action self-dual critical points of $\mathscr{A}_{\text{Yang Mills}}$ are known. These solutions generalize the one-instanton solution of

Belavin, Polyakov, Tyupkin and Schwartz,

$$A = \frac{x^2}{x^2 + \mu^2} \, gdg^{-1} \quad , \tag{2.4}$$

where for $G = SU(2)$, the map $g: \mathbb{R}^4 \to SU(2)$ is

$$g = |x|^{-1}(x^0 + ix^k\sigma^k) \quad .$$

Here $\{\sigma^k\}$, $k = 1,2,3$, are the Pauli matrices. The constant $\mu > 0$ is arbitrary, and is called the scale size. Generalizations of (2.4) are discussed in [42,27] and the most general solution was given in [7, 4, 17,13], by formulas similar in structure to (2.4).

The field A has a geometric interpretation as a connection on a principal bundle. If the connection A approaches the flat connection sufficiently rapidly as $|x| \to \infty$, then the integral

$$N = -\frac{1}{8\pi^2} \operatorname{Tr} \int F \wedge F \tag{2.5}$$

is an integer. A sufficient asymptotic condition is that A be the pullback of a connection on S^4 by the stereographic projection. This is because on S^4, the integral (2.5) is the second Chern number, which is invariant under pullback. It is a fact that all instanton solutions have integer N. In the example (2.4), $N = 1$.

In general the solution manifold for fixed N has dimension determined by N and the group G. For example with $G = SU(2)$, there is an $8|N| - 3$ parameter family of solutions [39, 5]. The instanton solutions minimize the action functional $\mathscr{A}_{\text{Yang Mills}}$, restricted to those connections for which N is a fixed integer. Further every local minimum of $\mathscr{A}$ is an instanton [9]. Thus while all minima of the Yang-Mills action are known, it is still open whether critical points of the action exist which are not minima. The expected answer is no, but only partial results are known [Chapter III §§10-11, 6, 9, 41].

In fewer than four dimensions, there are no finite-action solutions to the pure Yang-Mills equations (2.3); see Corollary II.2.3. The inclusion of the Higgs term in the action (1.8), results in a variety of additional phenomena. We summarize: For $d = 4$, the *only* Higgs solutions are finite-action solutions to the pure Yang-Mills equations (Corollary II.2.3). For $d > 4$, no finite action solutions at all exist to (1.17). However for $d = 2$ solitons do exist; they are called *vortices*. Solitons also exist for $d = 3$ where they are called *monopoles*. For both $d = 2$ and 3, these Higgs models display features qualitatively similar to the Yang-Mills theory in $d = 4$.

In particular, field configurations with sufficiently rapid decay define homotopy classes of maps of S^{d-1} into quotient manifolds G/J where $J \subseteq G$ is a subgroup of the gauge group. At least for particular values of λ and G ($\lambda = 1$, $G = U(1)$ in $d = 2$; $\lambda = 0$, G a simple Lie group in $d = 3$) the minima of $\mathscr{A}$ in certain of these homotopy classes exist and are solutions to the equations. Furthermore, for these values of λ the solutions satisfy first order equations analogous to the self-dual equations (2.1) for $d = 4$ Yang-Mills theories. The dimension of the solution set is determined by the group G and the homotopy class. For λ not equal to these special values, less is known about the answer to the existence question. In the later chapters we give *a priori* estimates on solutions to (1.17) for all values of $\lambda \geq 0$, in addition to the detailed study of the equations for the special values of λ above.

I.3. THE HIGGS EFFECT

The condition $|\Phi| \to 1$ as $|x| \to \infty$ gives the solutions to (1.17) characteristic masses, reflecting the fact that the Higgs model describes different particles: "vector particles" associated with the A field (photons) and "scalar particles" associated with the Φ field (mesons or Higgs particles). At first glance the action appears to describe massless vector particles, since no explicit mass enters $\mathscr{A}$. Furthermore, since each G has a $U(1)$ subgroup, we expect at least one massless meson. In fact massless particles occur only for certain field components, and massless mesons occur only for $\lambda = 0$.

Perhaps this terminology needs some explanation. Physicists associate to a field component with exponential decay $\exp(-m|x|)$ a particle of mass m, or a "length scale" m^{-1}. A particle of zero mass (infinite length scale) is associated to a field component which decays as a fundamental solution $u = |x|^{-d+2}$ to Laplace's equation, or as a derivative of u. Furthermore, the decay rates generally occur as square roots of the eigenvalues of the mass matrix, i.e., the Hessian of the nonderivative terms in the density (3.1) of $\mathscr{A}$, evaluated at the minimum of $\mathscr{A}$. Mathematical statements and proofs occur in Chapters III, IV. (In quantum theory a length scale is a necessary but not a sufficient condition for a physical particle to exist. See [23] for more details.)

How does the appearance of mass come about? We argue heuristically as follows. For the action (1.8) to be finite, it is necessary that as $|x| \to \infty$, $|\Phi| \to 1$ and $|D_A\Phi| \to 0$. Assuming the asymptotic values are approached sufficiently rapidly, then outside a small cone of solid angle we can gauge transform Φ such that $\hat{\Phi} \equiv \Phi/|\Phi|$ is constant vector in $\mathbb{L}$. The action density in this gauge is

$$a = \frac{1}{2}|F_A|^2 + \frac{1}{2}\nabla_j|\Phi|\nabla_j|\Phi| + \frac{1}{2}|\Phi|^2|\rho(A)(\hat{\Phi})|^2 + \frac{\lambda}{8}(|\Phi|^2 - 1)^2 . \tag{3.1}$$

Here $\rho(A)$ is the representation of A acting on the vector space in which Φ takes its values. This is given by $\rho(A)(\hat{\Phi}) = [A,\hat{\Phi}]$ when the model is the nonabelian Higgs model, and by $\rho(A)(\hat{\Phi}) = -iA\hat{\Phi}$ when the gauge group is $U(1)$. Thus for the abelian Higgs model,

$$|\rho(A)(\hat{\Phi})|^2 = |A|^2 .$$

For the nonabelian model,

$$|\rho(A)(\hat{\Phi})|^2 = \sum_{a,b} m^{ab}A_j^a A_j^b$$

where the sum is over a orthonormal basis of the Lie algebra $\mathfrak{g}$, and $\{m^{ab}\}$, as we will see, is the mass matrix. The matrix m^{ab} is positive. Thus the nonderivative terms in $\mathscr{A}$ are

$$\frac{1}{2}|\Phi|^2 \sum_{a,b} m^{ab} A_j^a A_j^b + \frac{\lambda}{8}(|\Phi|^2 - 1)^2 \quad . \tag{3.2}$$

Hence the Hessian at the minimum $(A = 0, \ |\Phi| = 1)$ gives rise to

$$m_{\text{Higgs}} = \sqrt{\lambda}$$

and

$$m_{\text{photons}} = (\text{eigenvalues } m_{ab})^{1/2} \quad .$$

We remark that a zero eigenvalue implies a power law decay of the corresponding component. Thus the Goldstone Theorem (by which a broken, continuous symmetry yields a massless particle) does not apply for these gauge theories. This is the "Higgs effect."

Let μ_1^2 denote the smallest nonzero eigenvalue of m^{ab}. The classical solutions in Chapters III-V exhibit asymptotic decay consistent with the Higgs effect and (for $\lambda \neq 0$) have nonzero masses. We only prove, however, that the Higgs mass is bounded below by the smaller of $\sqrt{\lambda}$ and $2\mu_1$. For vortices and SU(2) monopoles, $\mu_1 = 1$. For vortices we prove $m_{\text{photon}} \geq 1$. For vortices and SU(2) monopoles, we prove

$$m_{\text{Higgs}} \geq \min(\sqrt{\lambda}, 2) \quad .$$

Certain evidence suggests equality above. We conjecture

$$m_{\text{Higgs}} = \min(\sqrt{\lambda}, 2\mu_1) \quad . \tag{3.3}$$

In fact, if we set $|\Phi| = 1 + \eta$ in (3.2) then the action density for η, ignoring terms of order η^3, is

$$\hat{a}(\eta) = \frac{1}{2} \nabla_j \eta \nabla_j \eta + \lambda \eta^2 + \eta \sum_{a,b} m^{ab} A^a A^b \quad .$$

The field η is an approximate solution to the equation

$$(-\Delta+\lambda)\eta = -\sum_{a,b} m^{ab} A^a A^b ,$$

at least for large $|x|$. Hence its decay is governed not just by $\sqrt{\lambda}$, but also by the decay of $\Sigma_{a,b}\, m^{ab} A^a A^b$, suggesting the conjecture (3.3). Under the assumption that the fields are rotationally symmetric, (3.3) has been established for some equations [34]. See Dunlop and Newman for a discussion of related phenomena in the statistical mechanics of rotators.

I.4 ATTRACTIVE AND REPULSIVE FORCES

A general feature of the soliton solutions in $d = 2,3$ is that $|F|$ and $1 - |\Phi|$ are close to their asymptotic values, $|F| = |D_A\Phi| = 1 - |\Phi| = 0$, except in bounded regions of space. Typically, these regions have linear dimension m^{-1}, where m is the smallest nonzero mass generated by the Higgs effect. Furthermore, when these regions are separated by distances $d >> m^{-1}$ the fields in any one region are generally close to a spherically symmetric "fundamental" solution. The fundamental solution for $d = 2$, $G = U(1)$ is a vortex; for $d = 3$, arbitrary G, it is a monopole. (We say generally to avoid the case in which the local solution is a multivortex or multimonopole, corresponding to the superposition of fundamental solutions.)

For these reasons, it is convenient to think of the fundamental *solitons* as Newtonian *particles*, and the general soliton as a many particle solution.

As we remarked in §2, each configuration on $\mathbb{R}^d$ defines a homotopy class in the homotopy group $\pi_{d-1}(G/J)$ where $J \subseteq G$ is a subgroup. In all the Yang-Mills-Higgs models we consider, these groups $\pi_{d-1}(G/J)$ are either the integers $\mathbb{Z}$ or sums of various $\mathbb{Z}$'s (Chapter II.3). Thus addition and subtraction of homotopy classes are defined. The homotopy classes of the multi-particle solutions discussed in later chapters are sums of the homotopy classes of the constituent fundamental solitons. This gives further justification for the particle terminology.

For every soliton with a given homotopy class, there exists a soliton with the inverse class, and this correspondence is one-to-one. Since the homotopy groups are sums of integers, we can arbitrarily define positive and negative elements for fundamental solutions. The solutions which correspond to positive elements of π_{d-1} are called particles, as before. Those which correspond to negative elements are called anti-particles. It is common to refer to the homotopy class of a soliton as its *charge*, so particles are said to have positive charge. In §5,6 we see why it is natural to call this charge "magnetic."

The interpretation of solitons as particles leads to the question: What are the interparticle forces? In order for a static solution to exist, either the interparticle forces must sum to zero, or else some other stabilizing influence must be present. For example: Suppose the constituent particles attract; we would expect the static solutions to correspond to the configuration in which all particles are found near the same point. On the other hand, if particles repel, boundary conditions on the walls of finite domain are necessary to stabilize a static configuration of particles in the interior. In fact, solutions always exist in the interior of a bounded domain, for any value of the Higgs coupling constant λ. But when the net force is repulsive, we expect that there are no multiparticle solutions to the equations in infinite volume, except perhaps when all particles are located at the same point. If such a solution exists, it would not be a minimum of the action; rather it would be unstable with respect to a perturbation which tended to separate the particles.

Let us now discuss forces in more detail. Suppose a solution $(A_{12}\Phi_{12}) \equiv (12)$ contains a distant pair of fundamental solitons, denoted 1 and 2. Define the potential energy function

$$V = \mathscr{A}(12) - \mathscr{A}(1) - \mathscr{A}(2) \quad .$$

Here V is a function of the separation, and its gradient is interpreted as the interparticle force.

In analogy with electromagnetism, the force is mediated by the vector field A and the scalar field Φ. The vector potential creates an attractive force between fundamental solitons of opposite charge, but a repulsive force between fundamental solitons of equal charge. On the other hand, the scalar potential gives rise to attractive forces between fundamental charges, like or opposite.

Thus the possibility of the forces balancing can occur only for like charges. Furthermore this can happen only for one critical value of the coupling constant $\lambda = \lambda_{cr}$. For monopoles, $\lambda_{cr} = 0$, and for $\lambda > 0$ the net forces are repulsive. For vortices $\lambda_{cr} = 1$. For $\lambda > \lambda_{cr}$ the net force is repulsive, while for $\lambda < \lambda_{cr}$, like charges are attractive. The forces between opposite charges are always attractive.

In the case of vortices, Jacobs and Rebbi have made an approximate computer study of the 2-vortex potential. They found the qualitative picture of Figure 4.1, in agreement with the discussion above.

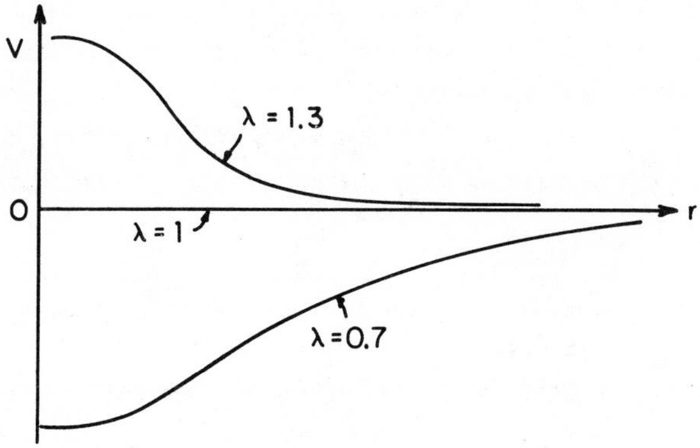

Figure 4.1. Two vortex potential (equal charge) for three values of the coupling constant λ.

For the critical values $\lambda_{cr}(\text{vortex}) = 1$, $\lambda_{cr}(\text{monopole}) = 0$ the interparticle potential vanishes; that is, solitons do not interact. In fact, multiparticle solutions corresponding to particles at arbitrary separations exist; cf. Chapters III and IV.

These solutions contain only particles or only antiparticles, which is also consistent with the above picture. Remark that even though the above argument rules out static solutions containing a particle and an antiparticle, we conjecture that time-dependent bound states do occur. In this case the constituents may bind into a "molecule," with rotational motion stabilizing the attractive forces between particle and antiparticle.

I.5. SUPERCONDUCTIVITY AND VORTICES

The solitons of the abelian Higgs model (vortices) occur in nature. For certain kinds of superconductors, the equilibrium state of the system is a multivortex configuration. We explain below.

Many materials when cooled below a critical temperature T_c, typically a few degrees Kelvin, exhibit the phenomena of superconductivity. The superconducting state is characterized by three macroscopic properties. First, electric currents flow without resistance. Second, magnetic fields vanish inside the superconducting medium; this phenomenon is called "flux expulsion" and is also known as the Meissner effect. Third, no net energy is released in the transition from the normal state to the superconducting state.

The superconducting state and the transition can be discussed in the framework of equilibrium statistical mechanics. Thus to a good approximation, the equilibrium state of the superconductor minimizes the free energy.

On the microscopic level, the superconductor is described by the theory of Bardeen, Cooper, and Schrieffer (BCS). In this theory, the onset of superconductivity is due to the formation of bound electron pairs (Cooper pairs). With respect to small applied forces, the electron pairs interact as single entity, a particle with twice the charge of a single electron. Furthermore this effective particle is a boson, since the paired state is a spin singlet. Large external forces disrupt the pairing and force the superconductor to the normal state. The superconducting state is characterized by an ensemble of Cooper pairs, and these in turn are described by a scalar field Φ for charged, spinless particles. As usual, such a field is a complex-valued function of position. The density $|\Phi(x)|^2$ is proportional to the number density of Cooper pairs. The field Φ is called the order parameter. The state where $\Phi \sim 0$ has few pairs and behaves as a normal conductor. The cases in which $|\Phi|$ is bounded away from zero describe the state of condensed pairs and superconductivity. The order parameter Φ provides a macroscopic description of the system described microscopically by the BCS theory. While this macroscopic theory can be obtained as an approximate consequence of the more general BCS theory, it was proposed well before the justification of a microscopic theory existed.

A system of equations for Φ and its interaction with the electromagnetic potential A was proposed by Ginzburg and Landau. It is based on thermodynamic considerations, and is an extension of Landau's theory of second order phase transitions. One assumes that the free energy density has an expansion in $|\Phi|$ and its derivatives,

$$a = a_0 + a_2|\Phi|^2 + a_4|\Phi|^4 + a_1|(-i\hbar\nabla - eA/c)\Phi|^2 \quad . \tag{5.1}$$

Here $\hbar$ is Planck's constant, c is the velocity of light and $e = 2e_{\text{electron}}$ is the electric charge of a Cooper pair. Also by assumption, the free energy a is an even function of Φ. The gradient term is the standard minimal coupling of an electromagnetic potential to a scalar field. The coefficients a_j are constants which depend on the temperature, the composition of the material, etc.

To ensure stability, $a_1, a_4 > 0$. We also assume $a_2 < 0$. In fact, if $a_2 > 0$, it follows from the maximum principle (Proposition VI.3.1) that a solution to the variational equations of (5.1) with $\int a < \infty$ must satisfy $\Phi \equiv 0$. Such a solution would describe the normal state.

After a scale change and the addition of a constant, we write (5.1) as

$$a = \frac{1}{2}|\nabla_A\Phi|^2 + \frac{\lambda}{8}(|\Phi|^2 - 1)^2 \quad . \tag{5.2}$$

Writing the free energy in the form (5.2) assumes that the magnetic potential A is given. In order to give a self-consistent description of the internal state of the superconductor, we must also regard A as a dynamical variable. Hence we add to (5.2) the free energy density of the magnetic field, $\frac{1}{2}|F_A|^2$. This yields the Ginzburg-Landau free energy

$$a = \frac{1}{2}|F_A|^2 + \frac{1}{2}|\nabla_A\Phi|^2 + \frac{\lambda}{8}(|\Phi|^2 - 1)^2 \quad . \tag{5.3}$$

However, (5.3) is just the action density of the abelian Higgs model in $d = 3$, namely (1.8) with $G = U(1)$, with $\mathbb{L} = \mathbb{C}$ and where ρ acts by multiplication. An especially relevant class of configurations is that for which the fields are approximately constant in one direction, say the x_3 direction. This assumption, and the gauge $A_3 = 0$, reduces (5.3) to the $d = 2$ abelian Higgs action.

We now concentrate on the variational equations of (5.3) to describe (A,Φ) inside a superconductor. The potential function $V(\Phi) = \lambda/8(|\Phi|^2-1)^2$ in (5.3) has a W shape, as illustrated in Figure 5.1.

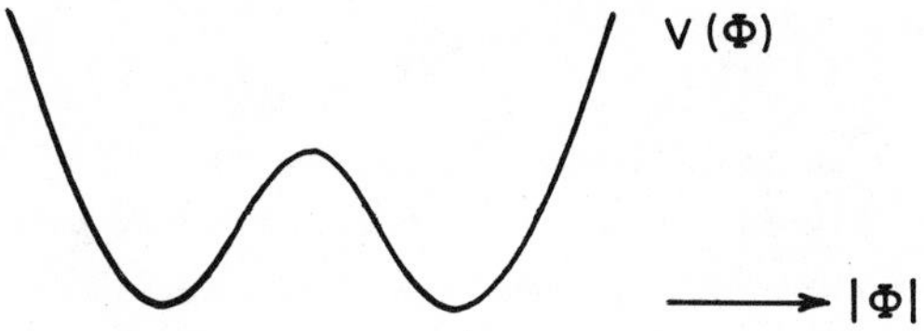

Figure 5.1. The Ginzburg-Landau potential function.

With minima which have been scaled to $|\Phi| = 1$ the state with $|\Phi| \sim 1$ is superconducting, while $|\Phi| \sim 0$ is normal.

Two qualitatively different types of superconductors are observed in nature, called types I and II. As we now explain, $\lambda < 1$ describes type I superconductors, while $\lambda > 1$ describes type II. The contrast between their properties is a consequence of the vortex-vortex force: whether it is attractive or repulsive.

A type I superconductor is characterized by complete expulsion of magnetic flux in the superconducting state. In fact for $\lambda < 1$, a mathematical bound shows that the magnitude of the magnetic field is bounded by the order parameter,

$$|F| \leq 1 - |\Phi| \qquad ;$$

see Chapter III. Hence flux is expelled in regions where $|\Phi| \sim 1$.

The Meissner effect in a type I superconductor is shown in Figure 5.2. A plot of the field strength $|F|$ inside the sample vs. the applied magnetic field H demonstrates a dramatic discontinuity at $|H| = H_{cr}$.

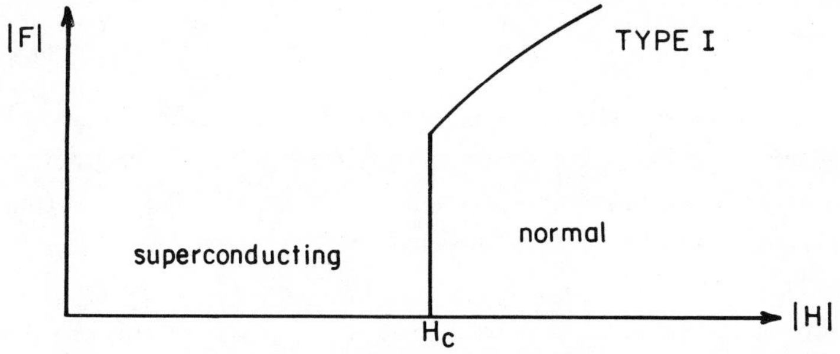

Figure 5.2. The Meissner effect in a type I superconductor.

On the other hand, a type II superconductor exhibits two critical fields H. For $H_{c1} < H < H_{c2}$, the magnetic flux penetrates the superconductor gradually (Figure 5.3.). On closer inspection, the curve of

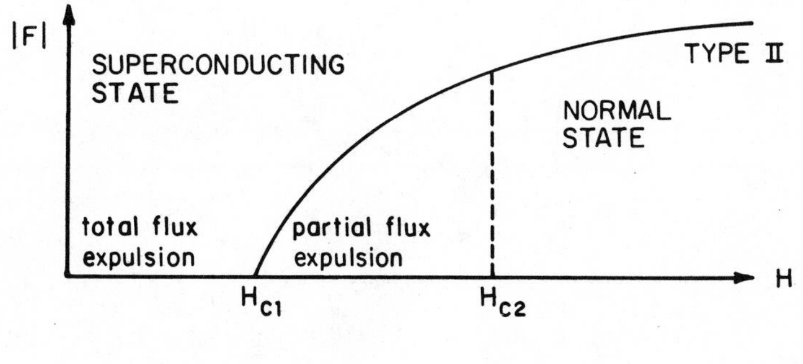

Figure 5.3. The superconducting transition for a type II material.

Figure 1.3 is not continuous for $H_{c1} < H < H_{c2}$. Its discontinuities arise from the fact that the flux penetrates the superconductor in tubes whose two-dimensional cross-section is a vortex. The flux carried in each tube is

$$\frac{hc}{e} = 2.07 \times 10^{-7} \text{ gauss cm}^{-2} \quad . \tag{5.4}$$

In the units in which the action density takes the form (5.3), each tube carries a magnetic flux 2π. Thus the region $H_{c1} < H < H_{c2}$ in a type II superconductor is characterized by a series of small increments. The total flux is an integer multiple of hc/e;

$$\int F = 2\pi N \tag{5.5}$$

in the units of (5.3).

The diameter of each flux tube found in experiments is of order 1 when expressed in the units of (5.3). This is precisely the length scale defined by the mass of the photon; and $m_{photon} = 1$. This length scale is called the penetration depth.

Figure 5.4 pictures the order parameter in the vicinity of the (cross-section) of a flux tube. It illustrates the vortex-like structure. On a length scale m^{-1}_{Higgs}, the magnitude of Φ becomes 1. This second length scale, defined by the Higgs mass, characterizes the spatial variation of the order parameter and is called the *correlation length.*

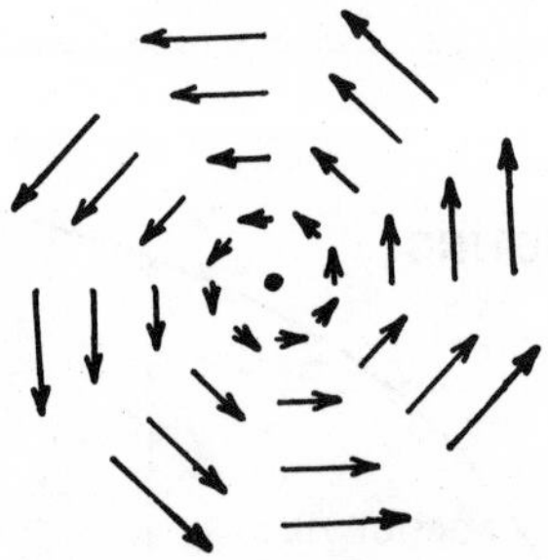

Figure 5.4. The order parameter Φ as a function of x, near a zero (vortex). We plot Φ as a vector with the $\text{Re}\,\Phi$ in the horizontal direction, $\text{Im}\,\Phi$ in the vertical direction.

Experiments deal with finite superconductors. For type II materials the flux tubes arrange themselves in a regular lattice because the vortex-vortex force is repulsive. The lattice spacing, i.e., the distance between vortex centers, is determined by the geometry of the superconductor and the total flux (5.5). The integer N is the vortex number. The significance of N as a winding number for the map Φ is treated in detail in Chapter II.3.

For fixed N the $\lambda > 1$ vortices are arranged in a lattice that maximizes the vortex-vortex separation. This lattice was predicted by Abrikosov and can be observed; it is illustrated in Figure 5.5. The square in the lower right of Figure 5.5 represents the area pictured in Figure 5.4.

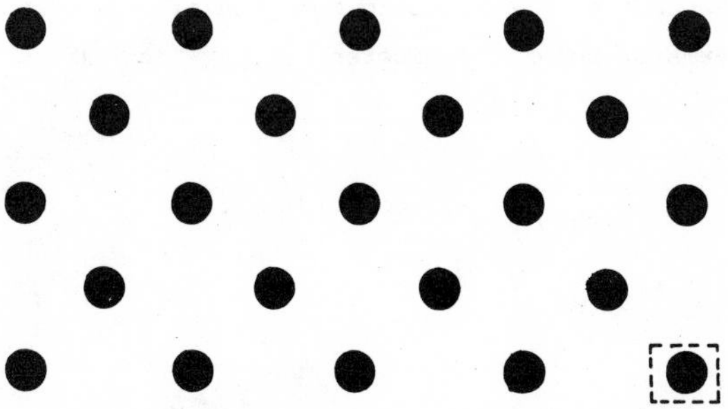

Figure 5.5. Regular lattice of type II vortices. See e.g. [28].

I.6. MONOPOLES

Dirac proposed the existence of a magnetic monopole on the basis of a singular solution to the Maxwell equations. Consider the Yang-Mills-Higgs variational equations (1.17) in the case that $d = 3$, $G = U(1)$, $\mathbb{L} = \mathfrak{g}$, ρ is the adjoint representation and $\lambda = 0$. The potential A is called the magnetic potential and Φ is the electric potential. The magnetic field is $*F = *dA$. Then (1.17) is the set of time-independent Maxwell equations in the absence of charges, namely

$$d{*F} = 0, \qquad dF = 0, \qquad \Delta\Phi = 0 \quad . \tag{6.1}$$

The only solution to (6.1) vanishing at infinity is $F = 0$, $\Phi = 0$.

Let us now introduce a density $q_e(x)$ of electric charge and a density $q_m(x)$ of magnetic charge. In their presence the static Maxwell equations are

$$d{*}f = 0 \quad , \tag{6.2a}$$

$$dF = (4\pi){*}q_m \quad , \tag{6.2b}$$

$$-\Delta\Phi = 4\pi q_e \quad . \tag{6.2c}$$

In the normal study of electrostatics the magnetic charge $q_m = 0$. Dirac's modification was to consider the case $q_e = 0$ and $q_m = \delta$, a delta function corresponding to a magnetic monopole of unit charge at the origin. In terms of polar coordinates (r,θ,χ) on $\mathbb{R}^3$, a solution to these equations is $\Phi \equiv 0$ and

$$A_{Dirac} = (1 - \cos\theta)d\chi \quad , \tag{6.3}$$

$${*}F_{Dirac} = \frac{1}{r^2}\, dr \quad . \tag{6.4}$$

As can be seen from the solution, the magnetic field is smooth except at $r = 0$ and has infinite energy, i.e., $F \notin L_2$. Considered as an $\mathbb{R}^3$ solution, the action is infinite. As is well known, the Dirac monopole has not been observed in nature. All magnets have two poles.

The Dirac monopole defines a connection on $\mathbb{R}^3 \setminus \{0\}$, which is topologically $S^2 \times \mathbb{R}$. On $S^2 \times \mathbb{R}$, the field F is a closed 2-form which is not exact. Its integral over a 2-sphere defines a magnetic charge whose value is one. Moreover, since the Maxwell equations are linear, any superpositions of monopoles are smooth infinite-action solutions on $\mathbb{R}^3 \setminus \{x_1,\dots,x_n\}$, where singularities occur at the monopole locations $\{x_1,\dots,x_n\}$.

A static electric charge at the origin corresponds, of course, to $q_m = 0$, $q_e(x) = e\delta(x)$. In this case, the solution to (6.2) is $F = 0$, $\Phi = e/r$. Thus Φ is singular at $r = 0$, and as in the magnetic case, the solution has infinite energy. One reason the infinite-energy electron solution is physically acceptable is that the point electric charge can be spread over a region. In other words the electron can be given an internal structure. In terms of equations, we replace $q_e(x) = e\delta(x)$ by a smooth, compactly-supported function of total integral e. In this case the electric potential Φ has finite energy.

On the other hand, in the case of the Dirac monopole, the charge density q_m cannot be a smooth distribution. If it were, then the equation $dF = *q_m$ implies that no solution to $F = dA$ exists. Thus the Maxwell equations (6.2) require modification to allow for finite-energy/action monopole solutions. For the same reasons we do not expect finite-action monopole solutions when $\lambda > 0$. (We call the $\lambda > 0$ equations the Maxwell-Higgs model.) We do expect an infinite-action monopole with a nonzero Higgs field Φ, and $1 - |\Phi| = O(1/r)$ as $r \to \infty$. In fact, all purely electromagnetic, i.e., $U(1)$, monopoles have infinite action.

Let us contrast this situation with the case of the *nonabelian* Yang-Mills-Higgs equations (1.17), to which *smooth, static finite-action* solutions do occur. These equations are central to the unified theory of weak and electromagnetic interactions as proposed by Glashow, Salam and Weinberg. From this point of view, the electromagnetic and weak interactions are two different manifestations of one force. This unification is made possible by regarding electromagnetic forces as arising from a $U(1)$ subgroup of the gauge group $G = SU(2) \times U(1)$. The theory predicts vector particles corresponding to the various components of the connection. Due to the Higgs effect, most of these are massive and mediate the short-distance weak interactions. On the other hand, at long distances the connection takes its values in a $U(1)$ subgroup (with exponentially small error). The theory as $r \to \infty$ is asymptotically the $U(1)$ Maxwell-Higgs theory, whose photon has zero mass and a characteristic $1/r$ decay.

At short distances, the Maxwell equation (6.2b) is replaced in the Yang-Mills-Higgs theory by the Binachi identity $DF = dF + [A,F] = 0$. Thus the commutator term acts as a smoothed out magnetic charge density term in the equation, even though an explicit charge $q_m(x)$ has not been introduced. For this reason we call these solutions magnetic monopoles; their charge, however, has topological significance and is a homotopy invariant (Chapter II.3). The solutions are asymptotically the Maxwell-Higgs monopoles, lying in the $U(1)$ subgroup of electromagnetism. (Asymptotically, these solutions differ from Dirac monopoles only in having a long range Φ field as well as a long range F field.) For short distances, however, the fields are inherently nonabelian, which is why the solution has finite action, and why it is smooth. For $\lambda = 0$, a solution in closed form was discovered by Prasad and Sommerfield; it is given in equation (IV.1.15). One expects that such magnetic monopoles should occur in nature, although they have yet to be seen.

The grand unified theories (GUTS) attempt to combine the $SU(2) \times U(1)$ model of weak and electromagnetic interactions with the $SU(3)$ theory of quantum chromodynamics (strong interactions). GUTS postulate that the groups involved are subgroups of some larger Lie group, such as $SU(5)$ or E_6. See [21,19]. Masses for the different subgroups are generated by the Higgs effect, and this is used to explain the strengths of various forces at different length scales. Monopole solutions for these larger gauge groups have possible significance in the theory of elementary particles [31] as well as cosmological implications [37,26]. For an extensive list of references, see [32].

I.7. CLASSICAL CONFIGURATIONS AND QUANTUM THEORY

The subject of quantizing a classical field theory is beyond the scope of this work, and we refer the reader to [23]. We explain in outline here why classical solutions are helpful in understanding quantum theory, and provide the leading terms in an approximation to quantum theory.

The main step in quantization of a classical action functional $\mathscr{A}$, defined on a space X of classical fields, is to construct a measure $d\mu$ on X. The space X is called the space of paths, in analogy to quantum mechanics. Basically $d\mu$ has a density $\exp(-\mathscr{A})$, and is normalized so that $\mu(X) = 1$. The moments of the measure μ are the expectations, analytically continued to imaginary time, of quantum fields in the vacuum state:

$$\int F(x_1)\ldots F(x_n)\Phi(y_1)\ldots\Phi(y_m)\,d\mu = \langle\Omega, F(x_1)\ldots F(x_n)\Phi(y_1)\ldots(y_m)\Omega\rangle. \tag{7.1}$$

Here Ω is the vacuum state and $<,>$ denotes the inner product on the Hilbert space of states. These moments yield, in principle, all spectral information about the Hamiltonian H, or more generally about the energy-momentum spectrum, including particles, bound states, scattering, resonances, etc.

Unfortunately, not much is known mathematically about the measures $d\mu$ for gauge theories. For non-gauge theories such measures have been constructed for $d=2,3$ [23]. In the case of gauge theories, the only example so far is the abelian Higgs model for $d=2$ [11]. In the case of scalar fields for $d=2,3$, the analysis has been carried to the point where it is clear what role the classical solutions play in understanding $d\mu$. We explain in outline how this is done, hoping that it serves as a guide for future work on these questions for gauge theories.

Let us then restrict attention to a scalar field Φ with action

$$\mathscr{A}(\Phi) = \frac{1}{2}\int|\nabla\Phi|^2\,dx + \int V(\Phi)\,dx \quad . \tag{7.2}$$

One property of the measure $d\mu$ is that the space $X_0 \subset X$ of finite-action fields Φ has measure zero. The local minima of $\mathscr{A}$ occur in $X_E \subset X_0$, and hence also have measure zero. Generically, the measure $d\mu$ is defined on a space of distribution-valued classical fields. For scalar fields, the space $X = \mathscr{S}'$ of Schwartz distributions is appropriate. Thus we are faced with

$$\int_{X_E} d\mu = \int_{X_0} d\mu = 0 \ , \qquad \int_X d\mu = 1 \quad . \tag{7.3}$$

So why are finite action configurations X_0 or classical solutions X_E useful in understanding $d\mu$?

The answer is that a point $\Phi_{c\ell}$ in field space can be the useful starting point for an approximation. Let us write $\Phi = \Phi_{c\ell} + \delta\Phi$, defining the fluctuation field $\delta\Phi$. Then Using Taylor's formula

$$\mathscr{A}(\Phi) = \mathscr{A}(\Phi_{c\ell}) + Q(\delta\Phi) + R(\delta\Phi) \quad . \tag{7.4}$$

Here Q contains terms linear and quadratic in $\delta\Phi$, while R denotes the remainder $\mathscr{A}(\Phi) - \mathscr{A}(\Phi_{c\ell}) - Q$. Let $\mathscr{A}_{c\ell} \equiv \mathscr{A}(\Phi_{c\ell})$ be the action of $\Phi_{c\ell}$, a number. Then

$$e^{-\mathscr{A}} = e^{-\mathscr{A}_{c\ell}} e^{-Q(\delta\Phi)} e^{-R(\delta\Phi)} \tag{7.5}$$

is a product of three factors:

$$\text{(classical)} \times \text{(Gaussian)} \times \text{(Remainder)}.$$

The success of using the decomposition (7.5) rests on whether the remainder factor $\exp(-R)$ is close to 1. We cannot expect this for all fields Φ, but only when $\delta\Phi$ is sufficiently small. The terms $\exp(-R)$ provide corrections to the Gaussian (semi-classical) approximation. Leading-order behavior can be exhibited by integration by parts with respect to the Gaussian measure with density $\exp(-Q)$. After integration by parts we evaluate the errors.

Thus we wish to find a set $\{\Phi_i\}$ of classical configurations with the following three properties: Each Φ_i is associated with an expansion

$$\mathscr{A}(\Phi_i) = \mathscr{A}_{c\ell,i} + Q_i(\delta\Phi_i) + R_i(\delta\Phi_i)$$

and a subset $\mathscr{R}_i \subset X$ of paths such that

$$\begin{array}{lll} \text{(i)} & \Phi \in \mathscr{R}_i \Rightarrow |e^{-R_i} - 1| & \text{small} \\ \text{(ii)} & \Phi \notin \mathscr{R}_i \Rightarrow e^{-Q_i} & \text{small} \\ \text{(iii)} & \cup \mathscr{R}_i = X, \quad \mathscr{R}_i \cap \mathscr{R}_j = \emptyset , & i \neq j \quad . \end{array} \tag{7.6}$$

The restriction that the $\mathscr{R}_i$ are nonoverlapping is merely a technical convenience. We write a partition of unity

$$1 = \sum_i \chi_i(\Phi) \quad , \qquad d\mu = \left(\sum_i \chi_i \, d\mu\right) \tag{7.7}$$

where

$$\chi_i(\Phi) = \begin{cases} 1 & \Phi \in \mathscr{R}_i \\ 0 & \Phi \notin \mathscr{R}_i \end{cases} .$$

The semiclassical approximation to an integral of a function $H(\Phi)$ is

$$\left(\int H(\Phi)\, d\mu\right)_{\text{semiclassical}} = \sum_i e^{-\mathscr{A}_{c\ell,i}} \int H(\Phi)\, d\nu_i \ . \tag{7.8}$$

where $d\nu_i$ is a Gaussian measure with weight $\exp(-Q_i)$. The error terms

$$\int H\, d\mu - \left(\int H\, d\mu\right)_{\text{semiclassical}} \tag{7.9}$$

have been proved to be small in the models mentioned at the start of the section. They arise from two sources. For $\Phi \in \mathscr{R}_i$ we have replaced $\exp(-R_i)$ by 1. Thus error terms of the form

$$0\left(\sum_i \sup_{\Phi \in \mathscr{R}_i} |e^{-R_i} - 1|\right)$$

occur. These are small by (i). Secondly, the integrals of each Gaussian over $X \setminus \mathscr{R}_i$ contribute errors

$$0\left(\sum_i \int_{X \setminus \mathscr{R}_i} |H(\Phi)|\, e^{-Q_i}\, d\mu\right) \quad ,$$

which are small by (ii). These errors occur because the characteristic functions χ_i of the partition of unity (7.7) have not been included in (7.8).

Why are classical solutions useful? First they provide a possible set of fields Φ_i for an expansion. Secondly, they can be used to construct Φ_i which are superpositions of exact solutions, i.e., approximate solutions, but which are not themselves solutions. For example, while we construct N-vortex solutions to the Higgs model, and while we show that vortex-antivortex solutions do not exist, it is presumably the case that vortex-antivortex superpositions are necessary for a detailed understanding of the quantized Higgs model.

The reason that solutions by themselves are insufficient to analyze the situation is that we can imagine the average field in different spatial regions taking on a configuration close to different solutions. Thus we require superpositions which interpolate between different minima. This is illustrated in the next section in the case of the Higgs model. In that case, the factors

$$e^{-\mathscr{A}_{c\ell,i}}$$

which arise from the classical terms can be interpreted, as i varies, as yielding a statistical mechanics of solitons. The energy of interaction between the classical solitons (particles) is described in §3. Thus we obtain a gas of particles with a statistical weight reflecting their interaction. In the case of the model above, a complete mathematical analysis of this picture is given in [24]. See also [10,30,35]. In the case of a vortex-antivortex pair, the potential is an attractive Yukawa potential. For a qualitative description of the statistical mechanics of this system, see [12,14]. In the monopole case, the qualitative features were worked out by [27,35]. For SU(2) monopoles, the long range forces in the statistical mechanics of a dilute gas should be Coulombic. Hence the screening analysis of Brydges and Federbush should be relevant. The classical solitons of earlier sections are generally believed to have an interpretation in quantum theory as particles.

II. TOPOLOGICAL ASPECTS

II.1 GEOMETRIC FIELD EQUATIONS: PRELIMINARIES

In the language of differential geometers, the gauge potential of Chapter I is a connection A on the principal G-bundle $P = \mathbb{R}^d \times G$, where G is a compact matrix group with Lie algebra $\mathfrak{g}$. The Higgs field Φ is a cross-section of the vector bundle $E = \mathbb{R}^d \times \mathbb{L}$ associated to P by the representation ρ of G on $\mathbb{L}$. The Yang-Mills-Higgs action functional $\mathscr{A}(A,\Phi)$ of (I.1.8) is defined on configurations (A,Φ) which satisfy a local regularity condition and which satisfy $|\Phi| \to 1$, $|D_A\Phi| \to 0$ at infinity sufficiently rapidly. For $d \leq 4$, a sufficient local regularity condition is that each component of the connection and of the Higgs field is locally square integrable, with locally square integrable first partial derivatives. As a consequence, a Sobolev inequality ensures that the action density is locally integrable. Unless otherwise stated, we require this local regularity, saying in the language of Chapters V.2 and VI.1, 2 that $(A,\Phi) \in L^1_{2,loc}$. This condition is invariant with respect to gauge transformations $g \in L^2_{2,loc}$.

Let T denote the tangent bundle to $\mathbb{R}^d$ and T^* the cotangent bundle. In a fixed gauge, A is a $\mathfrak{g}$-valued section of T^*. But it is important to distinguish between a connection and a Lie algebra valued 1-form. The space of connections is an affine space. By a Lie algebra valued 1-form, on the other hand, we mean a section of $\mathbb{R}^d \times (T^* \otimes \mathfrak{g})$ which is associated to the principal bundle P by the adjoint representation. This distinction is manifest in the way the two objects transform under gauge transformations. The connection transforms inhomogeneously, while the $\mathfrak{g}$-valued 1-form transforms under the adjoint action, as does the curvature F_A (see (I.1.12)). The difference between two connections, however, *is* a $\mathfrak{g}$-valued 1-form. The above distinction is implicit throughout.

In this chapter we consider topological and geometric aspects of finite action configurations (A,Φ). We begin §2 by deriving a virial theorem which states, at a critical point of $\mathcal{A}$, what fraction of $\mathcal{A}$ arises from each term in (I.1.8). As a corollary, we eliminate the possibility of nontrivial critical points occurring for certain values of d, of λ or of the boundary conditions.

In Sections 3-5 the topological invariants are introduced. Every (A,Φ) with sufficiently rapid asymptotic decay defines a homotopy class in the set of maps from the $(d-1)$-sphere at infinity into a homogeneous space. These homotopy invariants are classified by a set of integers. Further, in special cases, the classes are calculable from integrals of the curvature.

In the final section we outline the procedure of *dimensional reduction*. A pure Yang-Mills field configuration, invariant under a space-time symmetry, is shown to define a Yang-Mills-Higgs configuration on the lower-dimensional space of orbits of that symmetry.

II.2. THE STRESS TENSOR

The Maxwell stress tensor T_{ij} for the action functional (I.1.8) has components given by

$$T_{ij} = (F_{ik},F_{jk}) - \frac{1}{4}\delta_{ij}(F_{k\ell},F_{k\ell}) + \left(\nabla_{A_i}\Phi,\nabla_{A_j}\Phi\right)$$

$$- \frac{1}{2}\delta_{ij}\left(\nabla_{A_k}\Phi,\nabla_{A_k}\Phi\right) - \frac{\lambda}{8}\delta_{ij}(|\Phi|^2-1)^2 \quad . \tag{2.1}$$

By virtue of the variational equations (I.1.17) for a critical point of $\mathcal{A}$, the stress tensor satisfies the identity

$$\nabla_j T_{ij} = 0 \quad , \tag{2.2}$$

namely $d{*}T = 0$. A useful consequence of (2.2) follows by integration.

PROPOSITION 2.1. *Let* (A,Φ) *be a finite action solution to the second order eauations* (I.1.17). *Then the stress tensor* (2.1) *satisfies*

$$\int T_{ik}\,dx = 0 \quad , \tag{2.3}$$

where $i, k = 1,2,\ldots,d$.

<u>Proof</u>. Formally multiply (2.2) by x_k, and integrate (by parts). To justify neglecting the boundary term in the integration, let $b_R(|x|) = b(|x|/R)$, where $0 \le b(|x|) \le 1$ is a C_0^∞ monotonically decreasing function, equal to one for $|x| \le 1$. Then multiplying (2.2) by $x_k b_R(x)$ and integrating yields

$$\int T_{ik}(x) b_R(x)\,dx = -\int x_k(\partial_j b_R)(x) T_{ij}(x)\,dx \quad . \tag{2.4}$$

Note for some $\kappa > 0$,

$$(1 - b_{\kappa R}(x)) b_R'(x) \equiv b_R'(x)$$

and

$$|x|\,|(\partial_j b_R)(x)| \le \text{const} \quad . \tag{2.5}$$

Thus

$$\left|\int T_{ik}(x) b_R(x)\,dx\right| \le \text{const} \sum_j \int (1 - b_{\kappa R}(x))\,|T_{ij}(x)|\,dx \quad . \tag{2.6}$$

But each component $|T_{ij}(x)|$ is bounded by the integrand of the action (I.1.8) and hence is integrable. Thus the right side of (2.6) tends to zero as $R \to \infty$. The left side, however is bounded by $\int |T_{ik}(x)|\,dx < \infty$ by the same argument. Hence by the dominated convergence theorem

$$\lim_{R\to\infty} \int T_{ik}(x) b_R(x)\,dx = \int T_{ik}(x)\,dx = 0 \quad . \tag{2.7}$$

As a corollary, we obtain an equipartition theorem relating terms in $\mathscr{A}$.

COROLLARY 2.2. *Let* (A,Φ) *be a finite action solution to* (I.1.17). *Then*

$$\left(2-\frac{d}{2}\right)\|F\|_{L_2}^2 = \left(\frac{d}{2}-1\right)\|D_A\Phi\|_{L_2}^2 + \frac{\lambda d}{8}\,\||\Phi|^2-1\|_{L_2}^2 . \quad (2.8)$$

Proof. Use $\Sigma_i \delta_{ii} = d$, and (2.1), summed over $i=j$.

Remark. The identity (2.8) also follows formally from assuming that $\mathscr{A}$ is invariant at a critical point under an infinitesimal version of the scale transformations:

$$A_k(x) \to \alpha A_k(\alpha x) \quad , \qquad \Phi(x) \to \Phi(\alpha x) \quad . \quad (2.9)$$

However, we give the above proof to avoid discussion of the domain of the infinitesimal generator of (2.9).

As another corollary, we can eliminate certain Yang-Mills-Higgs models as having only trivial solutions:

COROLLARY 2.3. *Let* (A,Φ) *be a smooth, finite action solution to the Yang-Mills-Higgs equations* (I.1.17) (YMH for short).

(a) *A pure Yang-Mills theory in* $d<4$ *has no solution but* $F\equiv 0$.

(b) *If* $|\Phi|\equiv 1$ *and* $d<4$, *then* $D_A\Phi\equiv 0$, $F\equiv 0$.

(c) *If* $d=2$ *and* $\lambda=0$, *the only solution is* $F\equiv 0$ $|\Phi|\equiv \text{const}$.

(d) *If* $d>4$, *there are no nontrivial finite action solutions.*

(e) *A* YMH *solution in* $d=4$ *is gauge equivalent to pure* YM.

Proof. Statements (a) and (d) follow from (2.8). Statement (e) follows from (2.8) plus a gauge transformation. Statement (b) follows from the same identity, since $\lambda=0$ ensures $F\equiv 0$. By a smooth gauge transformation, $A=0$. Since $d\Phi\in L_2$, $\Delta\Phi=0$, it follows that $\Phi=\text{const}$.

Hence they must both vanish. To establish (b), note by (I.1.17) that

$$\nabla_A^2\Phi = 0 \quad .$$

Hence

$$0 = \Delta|\Phi|^2 = 2((\nabla_A)_j\Phi,(\nabla_A)_j\Phi) + 2(\Phi,\nabla_A^2\Phi) = 2|D_A\Phi|^2 \quad .$$

Thus $\|D_A\Phi\|_{L_2} = 0$. By (2.8), $\|F\|_{L_2} = 0$.

II.3. CLASSIFICATION BY HOMOTOPY

We study the homotopy classes of maps from S^{d-1} into quotient manifolds of G. This notion was worked out in detail in [7], with other work by [2,14,20,4,19,12]. We follow Goddard, Nuyts and Olive. Let $\hat{x}$ denote a point on the sphere S^{d-1}. Let

$$\tilde{\Phi}(\hat{x}) \equiv \lim_{t\to\infty} \Phi(t\hat{x})$$

denote a map from the sphere S^{d-1} at infinity in $\mathbb{R}^d$ into $\mathbb{L}$. Furthermore, let $\Phi(p_0)$ denote the asymptotic value of Φ on the positive x_1 axis, $\Phi(p_0) = \tilde{\Phi}(1,0,\ldots)$. Then define the Lie subgroup J as

$$J = \left\{ g \in G:\ \rho(g)(\Phi(p_0)) = \Phi(p_0) \right\} \quad . \tag{3.1}$$

THEOREM 3.1. *Let* A *be a continuous connection on* $P = \mathbb{R}^d \times G$ *and* Φ *a* C^1 *section of* $E = \mathbb{R}^d \times \mathbb{L}$. *Assume that*

$$\lim_{R\to\infty}\ \sup_{|x|=R} \left|1 - |\Phi|\right| = 0 \quad , \tag{3.2a}$$

and that for some $\delta > 0$

$$|x|^{1+\delta}|D_A\Phi| \le \text{const} \quad . \tag{3.2b}$$

Then

(a) *There exists a gauge such that* $\tilde{\Phi}(\hat{x})$ *is a continuous map from* S^{d-1} *into* $\mathbb{L}$.

(b) *The configuration* (A,Φ) *defines a homotopy class* $[(A,\Phi)] \in \pi_{d-1}(G/J)$ (Definition 4.5).

(c) *The class* $[(A,\Phi)]$ *is invariant under* C^1 *gauge transformations.*

(d) *Suppose* (a,ϕ) *are respectively a* C^1 $\mathfrak{g}$*-valued* 1*-form and a* C^1 *section of* $\mathbb{R}^d \times \mathbb{L}$ *which satisfy*

$$\lim_{R\to\infty} \sup_{|x|=R} |\phi| = 0 = \lim_{R\to\infty} \sup_{|x|=R} |x||a| \quad . \tag{3.3}$$

Then

$$[(A+a, \Phi+\phi)] = [(A,\Phi)] \quad .$$

Condition (3.2) is a sufficient condition to define the homotopy class $[(A,\Phi)]$. A necessary and sufficient condition is unknown.

CONJECTURE 1. *The class* $[(A,\Phi)]$ *can be defined in a "weak" sense under the assumption of finite action only. (See Schoen and Yau.)*

COROLLARY 3.2. *Let* (A,Φ) *be as in Theorem* 3.1. *Then there exists on every contractible set* $U \subset S^{d-1}$ *containing* p_0 *a map* $g(\hat{x}): U \to G$ *such that,*

$$\tilde{\Phi}(\hat{x})\Big|_U = \rho(g(\hat{x}))\Phi(p_0) \quad . \tag{3.4}$$

Next we identify the homotopy class of (A,Φ) with a homotopy class in the fundamental group $\pi_{d-1}(S^{n-1})$, where n is the dimension of $\mathbb{L}$. Let

$$e_R(\hat{x}) \equiv \Phi/|\Phi|$$

restricted to $|x| = R$.

PROPOSITION 3.3. *Suppose that* G *acts transitively on the unit* S^{n-1} *sphere in* $\mathbb{L}$. *Let* Φ *be a continuous,* $\mathbb{L}$*-valued function satisfying* (3.2a). *Then there exists* $R_0 < \infty$ *such that for all* $R > R_0$, *the map* $e_R(\hat{x})$ *is continuous and defines a homotopy class* $[\Phi]$ *in* $\pi_{d-1}(S^{n-1})$. *The class* $[\Phi]$ *is independent of* $R > R_0$. *If, in addition,* (A,Φ) *satisfy the conditions of Theorem* 3.1, *then* G/J *is diffeomorphic to* S^{n-1} *and* $[(A,\Phi)] = [\Phi]$.

Before proving these results in §4 we give three examples.

EXAMPLE 3.4. *Let* $d = 2$, $G = U(1) = \{z \in \mathbb{C}: |z| = 1\}$ *and* $\mathbb{L} = \mathbb{C}$. *The* G *action on* C *is multiplication. This action is transitive on the unit circle. The subgroup* J *is the identity, so* $G/J = U(1) \simeq S^1$.

Let Φ *be a continuous complex-valued function satisfying* (3.2a). *Then by Proposition* 3.3, Φ *defines a homotopy class* $[\Phi] \in \pi_1(S^1) = \mathbf{Z}$. *This integer* $[\Phi] = N$ *is called the vortex number.*

PROPOSITION 3.5. *Let* Φ *be as in Example* 3.4 *and let* (A,Φ) *satisfy the conditions of Theorem* 3.1. *Then*

$$\frac{1}{2\pi} \lim_{R\to\infty} \int_{|x|\leq R} F_A = N \quad . \tag{3.5}$$

Note that as per the convention adopted in §I.1 *for this example,* F_A *and* A *are real in* (3.5).

EXAMPLE 3.6. *Let* $d = 3$ *and* $G = SU(2) \simeq S^3$. *Take* $\mathbb{L} = su(2)$ *with* G *acting by the adjoint representation. As in the previous example, this action is transitive on* $S^2 \subset \mathbb{L}$. *The group* $J = S^1$ *and* $G/J \simeq S^2$; *this fibration is the Hopf map.*

Let Φ *be a continuous* $su(2)$*-valued function satisfying* (3.2a). *Then by Proposition* 3.3, *the homotopy class* $[\Phi] \in \pi_2(S^2)$ *is defined. Note that* $\pi_2(S^2) = \mathbf{Z}$, *so the class* $[\Phi]$ *defines an integer* N, *which is called the monopole number.*

PROPOSITION 3.7. *Let* Φ *be as in Example* 3.6. *Suppose that* (A,Φ) *satisfy the conditions of Theorem* 3.1 *and in addition,* $\mathcal{A}(A,\Phi) < \infty$ *and* $(1-|\Phi|) \in L_p$ *for some* $2 \le p \le 6$. *Then*

$$\frac{1}{4\pi} \operatorname{Tr} \int D_A\Phi \wedge F_A = N \tag{3.6}$$

where Tr *is the normalized trace of* §1.1.

EXAMPLE 3.8. *Take* $d=3$ *and let* G *be compact, simple, and simply connected. Take* $\mathbb{L} = \mathfrak{g}$ *with* G *acting by the adjoint representation. Suppose that* (A,Φ) *satisfies the conditions of Theorem* 3.1. *Then* $[(A,\Phi)]$ *is defined. The group* J *is necessarily the direct product of a torus,* T', *and a simply connected Lie subgroup,* $G' \subset G$ [7]:

$$J = T' \times G' \quad . \tag{3.7}$$

We compute $\pi_2(G/J)$: There is an exact homotopy sequence [18]

$$\cdots \to \pi_2(G) \to \pi_2(G/J) \to \pi_1(J) \to \pi_1(G) \to \cdots \quad . \tag{3.8}$$

Both $\pi_2(G)$ and $\pi_1(G)$ are trivial [1]. Thus

$$\pi_2(G/J) \simeq \pi_1(J) \quad . \tag{3.9}$$

To compute $\pi_1(J)$, use the exact homotopy sequence

$$\cdots \to \pi_1(G') \to \pi_1(J) \to \pi_1(T') \to 0 \quad . \tag{3.10}$$

which is valid since G' is connected. Both $\pi_1(G')$ and $\pi_0(G')$ are trivial. Let ℓ be the rank of G and $\ell - r$ the dimension of T'. From (3.10),

$$\pi_1(J) \simeq \pi_1(T') \simeq Z^{\ell - r} \quad . \tag{3.11}$$

Summarizing the above discussion, the homotopy class of $[(A,\Phi)]$ in $\pi_2(G/J)$ is uniquely defined by specifying $\ell - r$ integers,

$\{n_a\}_{a=1}^{\ell-r}$ with r defined by J: $r = \text{rank } G'$.

CONJECTURE 2. *Let* (A,Φ) *be as in Example* 3.8. *Assume that* $(A,\Phi) < \infty$ *and* $(1 - |\Phi|) \in L_p$, *for some* $2 \leq p \leq 6$. *Then*

$$\lim_{R\to\infty} \frac{1}{4\pi} \operatorname{tr} \int_{|x|=R} \Phi^m F_A = \sum_{a=1}^{\ell-r} b_m^a n_a \quad , \quad 1 \leq m \leq \ell - r \ . \tag{3.12}$$

The constants $\{b_m^a\}_{a,m=1}^{\ell-r}$ *depend only on the groups* $J \subset G$.

We remark that under a much stricter set of asymptotic conditions on F_A and Φ, the integers $\{n_a\}$ can be characterized in terms of the eigenvalues of the matrix

$$\beta(\hat{x}) = \lim_{|x|\to\infty} |x| x^i \varepsilon^{ijk} F_A(x)_{jk} \quad . \tag{3.13}$$

We refer the reader to [7] .

The proofs of Propositions 3.5 and 3.7 are deferred to §5.

II.4 ASYMPTOTIC DECAY AND INTEGER INVARIANTS

We begin by establishing the relation between the decay properties and invariants quoted as Theorem 3.1.

Proof of Theorem 3.1. Let $(r,\hat{x})$ denote polar coordinates on $\mathbb{R}^d$. From the definition of $\nabla_A\Phi$ we have

$$\frac{\partial}{\partial r} \Phi + \rho(A_r)(\Phi) = \nabla_{A_r} \Phi \quad . \tag{4.1}$$

It is always possible to choose a gauge such that $A_r = 0$. (See Uhlenbeck [21].) With this choice of gauge

$$\Phi(r,\hat{x}) = \Phi(r_0,\hat{x}) + \int_{r_0}^{r} dt (\nabla_{A_r}\Phi)(t,\hat{x}) \quad . \tag{4.2}$$

By (3.2b),

$$|\Phi(r,\hat{x}) - \Phi(r_0,\hat{x})| \le \text{const } r_0^{-\delta} \quad , \qquad r > r_0 \ . \tag{4.3}$$

Thus the sequence $\{\Phi(m,\hat{x})\}_{m=1}^{\infty}$ for fixed $\hat{x} \in S^{d-1}$ is Cauchy. It converges to a unique element $\tilde{\Phi}(\hat{x}) \in \{v \in \mathbb{L}: |v| = 1\}$.

The function $\tilde{\Phi}$ maps S^{d-1} into the unit sphere in $\mathbb{L}$. To show that $\tilde{\Phi}(\hat{x})$ is continuous, we prove:

LEMMA 4.1. *Let* $d(\hat{x},\hat{y})$ *be the geodesic distance between points* $\hat{x}, \hat{y} \in S^{d-1}$. *Given* $\varepsilon > 0$ *there exists* $\mu > 0$ *such that* $d(\hat{x},\hat{y}) < \mu$ *implies*

$$|\tilde{\Phi}(\hat{x}) - \tilde{\Phi}(\hat{y})| < \varepsilon \quad . \tag{4.4}$$

Proof. From (4.2) and (4.3) we obtain for $r_0 \le r_1$,

$$|\tilde{\Phi}(\hat{x}) - \tilde{\Phi}(\hat{y})| \le |\Phi(r_0,\hat{x}) - \Phi(r_0,\hat{y})| + \int_{r_0}^{r_1} dt |(\nabla_{A_r}\Phi)(t,\hat{x}) - (\nabla_{A_r}\Phi)(t,\hat{y})| + \frac{c}{8} r_1^{-\delta} \quad . \tag{4.5}$$

Here, c is a constant. Choose $r_1 = (2c/\varepsilon\delta)^{1/\delta}$. Because $\nabla_A\Phi$ is continuous and the integration range in (4.5) is finite, there exists $\mu > 0$ such that if $d(\hat{x},\hat{y}) < \mu$ then

$$|\Phi(r_0,\hat{x}) - \Phi(r_0,\hat{y})| + \int_{r_0}^{r_1} dt \, |(\nabla_{A_r}\Phi)(t,\hat{x}) - (\nabla_{A_r}\Phi)(t,\hat{y})| < \frac{\varepsilon}{2} \ . \tag{4.6}$$

Thus $|\tilde{\Phi}(\hat{x}) - \tilde{\Phi}(\hat{y})| < \varepsilon$ as claimed.

Define the subgroup $J \subset G$ by (3.1).

PROPOSITION 4.2. *The subgroup* J *is a Lie group and the quotient* $t: G \to G/J$ *is a smooth fibration. Further, the induced map* $\rho_*: G/J \to M = \{\rho(g)(\Phi(p_0)) : g \in G\}$ *is a homeomorphism.*

Proof. The subgroup J is closed since the representation ρ is continuous. That the quotient is a smooth fibration; see e.g. Helgason. The map ρ_* is continuous since t and ρ are. By definition it is onto. It is 1-1. Since G/J is compact and $M \subset \mathbb{L}$ is Hausdorff, it is a homeomorphism.

Now define a map from $S^{d-1} \to G/J$ as follows: Let $\bar{p}_0$ be the antipodal point to p_0 on S^{d-1}. If $\hat{x} \neq \bar{p}_0$ there is a unique shortest geodesic curve $s(\hat{x})$ that connects $\hat{x}$ and p_0. It is convenient to take as coordinates on S^{d-1} $(\theta(\hat{x}), \chi(\hat{x}))$, where $\theta(\hat{x})$ is the geodesic distance along $s(\hat{x})$ from p_0 to $\hat{x}$ and $\chi(\hat{x})$ are coordinates in S^{d-2}. This is the geodesic normal coordinate system centered at p_0 (polar coordinates on S^2 if $d = 3$).

Define for $\hat{x} \neq p_0$

$$V_A(r,\hat{x}) = \rho\left(P \exp\left[r \int_0^{\theta(\hat{x})} d\bar{\theta} A_\theta(r,\bar{\theta},\chi(\hat{x}))\right]\right), \tag{4.7}$$

where P means to path order the exponential and A_θ is defined by decomposition $A = A_r dr + A_\theta d\theta + A_\chi d\chi$. For finite r, $V_A(r,\hat{x})$ maps $S^{d-1} \setminus \{\bar{p}_0\} \to \mathrm{Aut}\,\mathbb{L}$, the group of automorphisms of $\mathbb{L}$.

LEMMA 4.3. *Let* $V_A(r,\hat{x})$ *be defined by* (4.7). *Then*

$$\lim_{r\to\infty} V_A(r,\hat{x})\,\tilde{\Phi}(p_0) \equiv \tilde{V}_A(\hat{x})$$

is a continuous map of $S^{d-1} \setminus \bar{p}_0 \to G/J$ *where*

$$J = \{g \in G: \rho(g)(\Phi(p_0)) = \Phi(p_0)\} \quad . \tag{4.8}$$

Proof. Use the identity

$$\frac{\partial}{\partial\theta}\Phi + \rho(A_\theta)\Phi = \nabla_{A_\theta}\Phi \tag{4.9}$$

to solve for $\Phi(r,\hat{x})$. When $x \neq p_0$, we obtain,

$$\Phi(r,\hat{x}) = V_A(r,\hat{x})(\Phi(r,p_0)) - \int_0^{\theta(\hat{x})} d\bar{\theta}\,(V_A(r,\bar{\theta},\chi(\hat{x}))\left((\nabla_{A_\theta}\Phi)(r,\bar{\theta},\chi(\hat{x}))\right). \tag{4.10}$$

Thus, by (3.2b),

$$|\Phi(r,\hat{x}) - V_A(r,\hat{x})(\Phi(r,p_0))| \leq \text{const } r^{-\delta} \tag{4.11}$$

Next eliminate the r dependence of Φ on the left hand side of (4.11) by using the fact that

$$|\tilde{\Phi}(\hat{x}) - \Phi(r\hat{x})| \leq \text{const } r^{-\delta} \quad . \tag{4.12}$$

This follows from (4.3). The result is

$$|\tilde{\Phi}(\hat{x}) - V_A(r,\hat{x})\tilde{\Phi}(p_0)| \leq \text{const } r^{-\delta}, \qquad \hat{x} \neq \bar{p}_0 \quad . \tag{4.13}$$

Thus, by Proposition 4.2, for fixed $\hat{x} \neq \bar{p}_0$, a Cauchy sequence in G/J is defined by $\{V_A(m,\hat{x})(\tilde{\Phi}(p_0))\}_{m=1}^{\infty}$. That is,

$$\lim_{m\to\infty} \rho_*^{-1}(V_A(m,\hat{x})(\tilde{\Phi}(p_0))) \equiv \tilde{h}(\hat{x}) \in G/J \tag{4.14}$$

exists and defines a map from $S^{d-1} \setminus \{\bar{p}_0\}$ into G/J. It is continuous because ρ_* is a homeomorphism and $\tilde{\Phi}(\hat{x}) = \rho_*(\tilde{h}(\hat{x}))$ is continuous.

LEMMA 4.4. *The map* $\tilde{h}(\hat{x})$ *defined by* (4.14) *has a unique extension to a continuous map from* $S^{d-1} \to G/J$.

Proof. For each point $x \in S^{d-2}$ define

$$V_A(r,\bar{p}_0,x) = \rho\left(P \exp r \int_0^{\pi} d\bar{\theta} A_\theta(r,\bar{\theta},x)\right) . \tag{4.15}$$

From (4.10) and (4.13),

$$|\tilde{\Phi}(\bar{p}_0) - V_A(r,\bar{p}_0,x)\tilde{\Phi}(p_0)| \le \text{const } r^{-\delta} , \quad x \in S^{d-2} . \tag{4.16}$$

For fixed $x \in S^{d-2}$ the sequence $\{V_A(m,\bar{p}_0,x)\tilde{\Phi}(p_0)\}_{m=1}^{\infty}$ defines a Cauchy sequence $\{h(m,x,\bar{p}_0)\}_{m=1}^{\infty} \in G/J$. It follows from (4.16) and Proposition 4.2 that for any points $x, x' \in S^{d-2}$,

$$\lim_{m\to\infty} h(m,x,\bar{p}_0) = \lim_{m\to\infty} h(m,x',\bar{p}_0) \equiv \tilde{h}(\bar{p}_0) . \tag{4.17}$$

Further, because $\tilde{\Phi}$ is continuous at $\bar{p}_0$, so is $\tilde{h}$.

DEFINITION 4.5. *The class* $[(A,\Phi)]$ *of a connection and Higgs field* (A,Φ) *which satisfies the conditions of Theorem* 3.1 *is defined to be the homotopy class of the map* $\tilde{h}(\hat{x}): S^{d-1} \to G/J$ *as given by Lemma* 4.4.

We show that the class $[(A,\Phi)]$ is gauge invariant. In fact, because $\mathbb{R}^d$ is contractible, every map $g: \mathbb{R}^d \to G$ is homotopic to the identity. Thus $[(g^{-1}Ag + g^{-1}dg, g^{-1}\Phi g)]$ is homotopic to $[(A,\Phi)]$.

As for statement (d) of Theorem 3.1, it follows from (3.3) that

$$\lim_{r\to\infty} (\Phi + \phi)(r,\hat{x}) = \tilde{\Phi}(\hat{x}) . \tag{4.18}$$

Further, from (4.7),

$$\lim_{r\to\infty} V_{A+a}(r,\hat{x})(\Phi(p_0)) = \lim_{r\to\infty} V_A(r,\hat{x})(\Phi(p_0))$$

$$= \tilde{V}_A(\hat{x})(\Phi(p_0)) . \tag{4.19}$$

Therefore $[(A+a,\Phi+\phi)] = [(A,\Phi)]$.

Proof of Corollary 3.2. The corollary follows from the fact that the projection $t: G \to G/J$ is a fibration. The map $\tilde{\Phi}$ over U is homotopic to a point so by the homotopy lifting property of a fibration $\tilde{\Phi}$ lifts to a map $g(x/|x|): U \to G$ which commutes with the projection t (cf. Spanier [18]).

Proof of Proposition 3.3. Since $|\Phi| \to 1$ uniformly, there exists a radius $R_0 < \infty$ such that for all $R > R_0$, $|\Phi| > 1/2$. Hence for all $R > R_0$ the map $e_R: S^{d-1} \to S^{n-1} \subset \mathbb{L}$ is continuous. Further, e_R varies continuously with R for $R > R_0$. Hence the class of s_R in $\pi_{d-1}(S^{n-1})$ is independent of $R > R_0$. We denote this class by $[\Phi]$.

If (A,Φ) satisfies the conditions of Theorem 3.1 then $\tilde{\Phi}(p_0)$ is well-defined. The map $\rho_*: G/J \to S^{n-1}$ defined by

$$g \to \rho(g)(\Phi(p_0))$$

is a diffeormorphism between G/J and S^{n-1} (see Helgason). Thus $[\rho_*^{-1} e_R] \in \pi_{d-1}(G/J)$. By construction this is just $[(A,\Phi)]$.

II.5. INTEGRAL INVARIANTS AND THE CURVATURE

In the examples of §3, the integral homotopy invariants of Theorem 3.1 are expressible as curvature integrals. The proof of this fact follows.

Proof of Proposition 3.5. By assumption, there exists R_0 such that $|\Phi(x)| > 1/2$ if $|x| > R_0$. For $|x| > R_0$,

$$e(x) = \Phi(x)|\Phi(x)|^{-1} \tag{5.1}$$

is C^1. By definition,

$$D_A\Phi = ed|\Phi| + |\Phi|e(\bar{e}de - iA) ,$$

$$\overline{D_A\Phi} = \bar{e}d|\Phi| + |\Phi|\bar{e}(ed\bar{e} + iA) . \qquad (5.2)$$

Thus for $|x| > R_0$,

$$A + \frac{i}{2}(\bar{e}de - ed\bar{e}) = \frac{i}{2}(\bar{e}D_A\Phi - e\overline{D_A\Phi})|\Phi|^{-1}. \qquad (5.3)$$

The path integral of A along the circle $|x| = R$ is computed from (5.3):

$$\frac{1}{2\pi}\int_{|x|=R} A = -\frac{i}{4\pi}\int_{|x|=R}(\bar{e}de - ed\bar{e}) + \frac{i}{4\pi}\int_{|x|=R}(\bar{e}D_A\Phi - e\overline{D_A\Phi})|\Phi|^{-1} . \qquad (5.4)$$

The first term on the right hand side of (5.4) is the pullback $e^*(d\theta)$ of the generator of the first cohomology group $H^1(S^1)$. Here $e(R,\hat{x})$: $S^1 = \{x: |x| = R\} \to S^1 = \{z \in \mathbb{C}: |z| = 1\}$. By the Hurewicz isomorphism theorem [18], $H^1(S^1) \simeq \pi_1(S^1)$, and the first integral on the right side of (5.4) equals the vortex number N, independent of R. Thus from (3.2b), for $R > R_0$,

$$\left|\left(\frac{1}{2\pi}\int_{|x|=R} A\right) - N\right| \le cR^{-\delta} . \qquad (5.5)$$

Equation (3.5) follows by Stokes' Theorem. We remark that without assuming that $F_A \in L_1$, it is not correct to say that

$$\frac{1}{2\pi}\int_{\mathbb{R}^2} F_A = N .$$

<u>Proof of Proposition 3.7.</u> There exists by assumption $R_0 < \infty$ such that if $|x| > R_0$ then $|\Phi(x)| > 1/2$. For $|x| > R_0$, define

$$e = \Phi|\Phi|^{-1} \in C^1(\mathbb{R}^3 \setminus \{x: |x| < R_0\}) \quad . \tag{5.6}$$

Once again, the Hurewicz isomorphism theorem states that $\pi_2(S^2) \simeq H^2(S^2)$ and it allows us to calculate N by integrating $e^*(\omega)$ over the two sphere S_R^2 of radius $R > R_0$. Here $e^*(\omega)$ is the pullback of the volume two-form $\omega = *1$ on S^2. Explicitly for $R > R_0$,

$$N = \frac{1}{4\pi} \int_{|x|=R} (e, de \wedge de) = -\frac{1}{4\pi} \int_{|x|=R} \operatorname{Tr}\{ede \wedge de\} \tag{5.7}$$

where Tr is the normalized matrix trace (1.6).

PROPOSITION 5.1. *Let* Φ *be as in Example* 3.6, *and suppose that* (A,Φ) *satisfies the conditions of Theorem* 3.1. *Then*

$$\lim_{R\to\infty} \frac{1}{4\pi} \int_{|x|=R} |\Phi|^{-1}(\Phi, F_A) = N \quad . \tag{5.8}$$

Proof. From (5.7), and for $R > R_0$,

$$N = -\frac{1}{4\pi} \operatorname{Tr} \int_{|x|=R} \{eD_A e \wedge de - e[A,e] \wedge de\} \quad ,$$

$$= -\frac{1}{4\pi} \operatorname{Tr} \int_{|x|=R} \{eD_A e \wedge D_A e - e[A,e] \wedge de - ede \wedge [A,e] - e[A,e] \wedge [A,e]\} . \tag{5.9}$$

Now use the identities

$$e^2 = -\frac{1}{4}\mathbf{1} \, , \qquad ede = -dee \quad , \tag{5.10}$$

to obtain

$$N = -\frac{1}{4\pi} \operatorname{Tr} \int_{|x|=R} \{eD_A e \wedge D_A e + de \wedge A - eA \wedge A\} \, , \quad R > R_0 \quad .$$

Also, for $R \geq R_0$,

$$|D_A e| \leq 4|D_A \Phi| \leq cR^{-1-\delta} \quad . \tag{5.11}$$

Thus, integration by parts gives

$$\left| N + \frac{1}{4\pi} \int_{|x|=R} (e,F_A) \right| \leq cR^{-2\delta} \ , \qquad \text{for} \quad R > R_0 \quad .$$

The Proposition follows.

Next, define $b_R(x) \in C_0^\infty(|x| < 2R)$ as in VI.3.6. By Stokes' Theorem,

$$\int_{|x|=2R} (e,F_A) = -\operatorname{Tr} \int_{|x|\leq 2R} \Big\{ (1-b_R) D_A e \wedge F_A + b_R D_A \Phi \wedge F_A + (|\Phi| - 1) db_R \wedge eF_A \Big\} \quad . \tag{5.12}$$

By assumption, F_A, $D_A \Phi \in L_2$ and by (5.11), $D_A e \in L_2$. Thus by Hölder's inequality,

$$\left| \int_{|x|=2R} (e,F_A) + \operatorname{Tr} \int_{|x|\leq 2R} D_A \Phi \wedge F_A \right| \leq c_1 \|F_A\|_{L_2} \|(1-b_R) D_A \Phi\|_{L_2} + \|F_A\|_{L_2} \|\nabla b_R (1-|\Phi|)\|_{L_2} \quad . \tag{5.13}$$

We show that the right hand side of (5.13) goes to zero as $R \to \infty$. Note that $\nabla b_R = (1-b_{R/2}) \nabla b_R$ so

$$\| \nabla b_R (1-|\Phi|) \|_{L_2} \leq \|\nabla b_R\|_{L_{2p/(p-2)}} \|(1-b_{R/2})(1-|\Phi|)\|_{L_p}$$

$$\leq R^{(p-6)/2p} \|\nabla b\|_{L_{2p/(p-2)}} \|(1-b_{R/2})(1-|\Phi|)\|_{L_p} \quad . \tag{5.14}$$

By Proposition VI.1.3, the right hand side of (5.13) goes to zero as $R \to \infty$. Therefore

$$\lim_{R\to\infty} \frac{1}{4\pi} \operatorname{Tr} \int_{|x|\le R} \{D_A\Phi \wedge F_A\} = N \quad . \tag{5.15}$$

The assumption of finite action and the dominated convergence theorem gives (3.6).

II.6. SYMMETRY AND EQUATIONS

Consider a pure Euclidean Yang-Mills connection on $\mathbb{R}^d$ which is independent of one coordinate x^d. It is equivalent to a Yang-Mills-Higgs model on $\mathbb{R}^{d-1}$. In fact, since

$$F_{jd} = \delta_j A_d + [A_j, A_d] \quad ,$$

defining $\Phi = A_d$ and $D_A\Phi = d\Phi + [A,\Phi]$, we have

$$F_{jd} = (D_A\Phi)_j \quad .$$

Letting F denote the 2-form

$$F = \frac{1}{2} \sum_{i,j=1}^{d-1} F_{ij} dx^i \wedge dx^j \quad ,$$

the action per unit x^d is

$$\mathscr{A} = \frac{1}{2} \|F\|^2_{L_2(\mathbb{R}^{d-1})} + \frac{1}{2} \|D_A\Phi\|^2_{L_2(\mathbb{R}^{d-1})} \quad . \tag{6.1}$$

This the (d-1)-dimensional Yang-Mills-Higgs action with $\mathbb{L} = \mathfrak{g}$, and ρ the adjoint action.

Note that this example is different from the example in Chapter I.1 where we considered time-independent solutions to the Minkowski Yang-Mills-Higgs equations. In that case we required $A_0 = 0$ to obtain static solutions on (d+1)-dimensional Minkowski space.

Both these examples, however, illustrate a phenomenon known as dimensional reduction (see Forgács and Manton, or Mayer). The reduction of the pure Yang-Mills on $\mathbb{R}^d$ to (6.1) is the simplest example.

In the general case let M be a Riemannian manifold on which a connected Lie group S acts as a smooth symmetry group; i.e., there is a C^∞ map

$$\alpha\colon S\times M\to M \tag{6.2}$$

such that $\alpha(u_1u_2,m)\to\alpha(u_1,\alpha(u_2,m))$ for $u_1,u_2\in S$ and $m\in M$. Let $p\colon P\to M$ be a principal G-bundle over M and let $\bar{\alpha}\colon S\times P\to P$ be a lifting of the action (6.2) to P such that

$$\text{(1)}\quad p\circ\bar{\alpha} = \alpha\circ p \tag{6.3}$$

and

(2) for fixed $u\in S$, the map $\bar{\alpha}(u,\cdot)$ commutes with the right action of G on P. This means that $\bar{\alpha}(u,\cdot)\colon P\to P$ is a bundle map.

We define an S-invariant connection in the following way: Recall that a connection A on P defines a horizontal subspace $H_A\subset T_P$ by the condition that $H_A=\{\text{tangent vectors } \xi\in T_P \mid A(\xi)=0\}$. That is, A is a Lie algebra-valued 1-form on P so at each point $y\in P$ it maps the tangent space at y, $T_{Py}\to \mathfrak{g}$. The subspace H_A is the kernel of this map. The differential of $\bar{\alpha}(u,\cdot)$ is the linear map $\bar{\alpha}_*(u)\colon T_{Py}\to T_{P\bar{\alpha}(u,y)}$.

DEFINITION 6.1. *A connection on* P *is invariant with respect to the* S*-action* $\bar{\alpha}\colon S\times P\to P$ *iff the induced map* $\bar{\alpha}_*(u)\colon T_P\to T_P$ *maps* H_A *into* H_A *for all* $u\in S$. *The connection is said to be* S*-invariant in this case.*

The condition of invariance can be expressed directly as a differential condition on the 1-form A: Let $\mathfrak{s}$ denote the Lie algebra of S. For $s\in\mathfrak{s}$, the map $t\to\exp(ts)$ of $\mathbb{R}$ into S is a 1-parameter subgroup. Hence the map $\bar{\alpha}_s(t,\cdot) = \bar{\alpha}(\exp(ts),\cdot)$ is a one-parameter group of transformations of P whose generator is the vector field $\dot{\bar{\alpha}}_s = \frac{d}{dt}\,\bar{\alpha}(\exp(ts),\cdot)$.

PROPOSITION 6.2. *Definition* 6.1 *is equivalent to the statement that a connection* A *is* S-*invariant iff*

$$\mathscr{L}_{\bar{\alpha}_s} A = 0 \quad , \tag{6.4}$$

for all $s \in \mathcal{s}$. *Here* $\mathscr{L}$ *denotes the Lie derivative.*

Proof. To see the if, note that condition (6.4) implies that for all vectors $\xi \in T_{P_y}$, $y \in P$,

$$\frac{\partial}{\partial t} [A(\bar{\alpha}_s(t,y))\bar{\alpha}_{s*}(t,y)\xi] = 0 \quad . \tag{6.5}$$

If $\xi \in H_{A_y}$, (6.5) states that $\bar{\alpha}_{s*}(t,y)\xi \in H_{A_{\bar{\alpha}_s(t,y)}}$. Since S is generated by elements of the form $\exp(ts)$, $s \in \mathcal{s}$, the definition follows.

On the other hand, if Definition 6.1 is assumed, then for all $y \in P$ and horizontal vectors $\xi \in H_{A_y}$ equation (6.5) holds. Hence $(\mathscr{L}_{\bar{\alpha}_s} A)(\xi) = 0$ for all horizontal vectors ξ. As for the vertical subspace $V \subset T_P$, $\bar{\alpha}_{s*}$ maps vertical vectors into vertical vectors because of (6.3). In fact because $\bar{\alpha}_s$ is a bundle map (that is, it commutes with the action of G), $\bar{\alpha}_{s*}$ is the identity map on vertical vectors. This is seen as follows. Let $y \in P$. A vertical vector field is equivalent to a 1-parameter curve in P given by $e^{\varepsilon\eta}$, $\varepsilon \in \mathbb{R}$ and $\eta \in g$. Because $\bar{\alpha}_s$ is a bundle map,

$$\bar{\alpha}_s(t, ye^{\varepsilon\eta}) = \bar{\alpha}_s(t,y)e^{\varepsilon\eta} \quad . \tag{6.6}$$

Hence letting $\hat{\eta}$ represent the vector field induced by $e^{t\eta}$, we have $\bar{\alpha}_{s*}(\hat{\eta}|_y) = \hat{\eta}|_{\bar{\alpha}_s(y)}$. In addition,

$$A(\bar{\alpha}_s(y))\,\hat{\eta}|_{\bar{\alpha}_s(y)} = \eta = A(y)(\hat{\eta}|_y) \quad . \tag{6.7}$$

This last is one of the defining properties of A, cf. Kobayashi and Nomizu, [11], Chapter 2. This establishes (6.4) on vertical vectors. Since $T_P = V + H_A$, the Proposition follows.

To summarize, given the actions $\alpha\colon S\times M\to M$ and $\bar{\alpha}\colon S\times P\to P$, the problem of categorizing S-invariant connections reduces to the problem of finding the general solution to 6.4. The categorization of all S-invariant connections is due to Wang (cf. Kobayashi and Nomizu, §11).

Rather than discuss the abstract situation in greater detail we give a nontrivial example.

Let M be a 4-dimensional Riemannian manifold on which $SO(3)$ acts as a group of isometries. Without loss of generality we assume at least locally that the metric has the form

$$ds^2 = g_t(t,r)dt^2 + g_R(t,r)dr^2 + R^2(t,r)(d\theta^2 + \sin^2\theta d\chi^2) \quad . \tag{6.10}$$

For example, if $M=\mathbb{R}^4$ then $(g_t,g_r,R) = (1,1,r)$ and if $M=\mathbb{R}^2\times S^2$, $(g_t,g_r,R) = (1,1,\sqrt{2})$.

Denote the generators of $\mathfrak{so}(3)$ by $\{s^i\}_{i=1}^3$ with $[s_i,s_j] = -\varepsilon^{ijk}s_k$. Because the action of $SO(3)$ on M is a left action, the vector fields on M which are the infinitesimal generators of the $SO(3)$ action form a Lie algebra that is anti-isomorphic to $\mathfrak{so}(3)$. That is, set

$$\ell_i\big|_x = \dot{\alpha}_{s_i}(0,x) \;, \qquad x\in M \quad .$$

Then

$$[\ell_i,\ell_j] = \varepsilon^{ijk}\ell_k \quad . \tag{6.11}$$

The vectors $\{\ell_j\}_{j=1}^3$ are

$$\ell_1 = -\left(-\sin\chi\,\frac{\partial}{\partial\theta} - \cos\chi\,\cot\theta\,\frac{\partial}{\partial\chi}\right) \;,$$

$$\ell_2 = -\left(\cos\chi\,\frac{\partial}{\partial\theta} - \sin\chi\,\cot\theta\,\frac{\partial}{\partial\chi}\right) \;,$$

$$\ell_3 = -\frac{\partial}{\partial\chi} \quad . \tag{6.12}$$

Similarly, the infinitesimal generators $\{\bar{\ell}_i\}_{i=1}^3$ of the SO(3) action $\bar{\alpha}$ on P form a Lie algebra that is also anti-isomorphic to $\mathfrak{so}(3)$:

$$\bar{\ell}_i\Big|_y = \dot{\bar{\alpha}}_{s_i}(0,y) \quad , \qquad y \in P \quad ,$$

and

$$[\bar{\ell}_i,\bar{\ell}_j] = \varepsilon^{ijk}\,\bar{\ell}_k \quad . \tag{6.13}$$

Further the induced map $p_*: T_P \to T_M$ is a Lie algebra isomorphism of $\{\bar{\ell}_i\}$ with $\{\ell_i\}$.

A basis for the vertical vector fields on P is given by the infinitesimal generators of the automorphism induced by the right action of G,

$$y \to ye^{t(-i/2\,\sigma^j)} \quad , \qquad \text{for } t \in \mathbb{R}, \quad y \in P \tag{6.14}$$

and $\{-i/2\,\sigma_j\}_{j=1}^3$ a basis for $\mathfrak{su}(2)$. Let $\{e_j\}_{j=1}^3$ be the vector fields that generate (6.14) on P. Since (6.14) is a right action, the correspondence $-i/2\,\sigma_j \to e_j$ is an isomorphism of Lie algebras.

Let $x_0 \in M$ and suppose that x_0 is not a fixed point of the SO(3) action. Let B be a small open ball centered at x_0 such that B contains no fixed point under SO(3). With respect to a trivializatio of P over B (see Chapter V.4), there are two possible choices for the generators $\tilde{\ell}_j$. They are

$$(1) \qquad \tilde{\ell}_j = \ell_j \quad , \tag{6.15}$$

or

$$(2) \qquad \tilde{\ell}_j = \ell_j + e_j . \tag{6.16}$$

For case (1), the connection is trivial when restricted to any two-sphere. The more interesting case is (2). The Lie algebra valued 1-form that physicists generally associate with a connection is a 1-form on M that is pullback of A via a local section of P. Let $U \subseteq M$ and let $h: U \to P$ be a smooth section. The general form for a connection that is invariant with respect to transformations of P that are generated by (2) is [22, 5]

$$h^*A = hdh^{-1} + \frac{1}{2} h\ [aQ + (\Phi_1 - 1) QdQ + \Phi_2 dQ] h^{-1} \quad , \tag{6.18}$$

where $a = a_r(r,t)dr + a_t(r,t)dt$ is a real one-form on M and $\Phi_{1,2} = \Phi_{1,2}(r,t)$ are real-valued functions. Here

$$Q = i(\cos\theta\ \sigma^3 + \sin\theta(\cos\chi\ \sigma^1 + \sin\chi\ \sigma^2)) \quad . \tag{6.19}$$

The invariance of A follows from the fact that

$$\ell_j Q - \frac{i}{2}[\sigma^j, Q] = 0 \quad . \tag{6.20}$$

Modulo gauge rotations, the curvature is:

$$F_A = +\frac{1}{2} daQ + \frac{1}{2}(d\Phi_1 + a\Phi_2) \wedge QdQ + \frac{1}{2}(d\Phi_2 - a\Phi_1) \wedge dQ$$
$$- \frac{1}{4}(1 - |\Phi_1|^2 - |\Phi_2|^2) dQ \wedge dQ \quad . \tag{6.21}$$

Now set $F_a = da$, $\Phi = \Phi_1 + i\Phi_2$ and $D_a\Phi = d\Phi - ia\Phi$. The action for the Yang-Mills connection A is

$$\mathscr{A}_M(A) = \frac{1}{2}\int_M - 2\mathrm{tr}(F_A \wedge *F_A) = \frac{1}{2}\|F_A\|^2_{L_2} \quad .$$

The action density, $-2\mathrm{tr}(F_A \wedge *F_A)$ is by construction, constant on the 2-spheres $(r,t) = \mathrm{const}$. Hence the integration over these two-spheres is trivial:

$$\mathscr{A}_M(A) = \mathscr{A}_{M/SO(3)}(a,\Phi) = 2\pi \int_{M/SO(3)} R^2(r,t)\sqrt{g_t g_r}\ dr\ dt$$
$$\left\{ (g^t g^r)^{-1} |F_{art}|^2 + 2R^{-2} g_t^{-1} |\nabla_{a_t}\Phi|^2 + 2R^{-2} g_r^{-1} |\nabla_{a_r}\Phi|^2 \right.$$
$$\left. + \frac{1}{R^4}(1 - |\Phi|^2)^2 \right\} \quad . \tag{6.22}$$

Further, the quadratic curvature invariant (second Chern class)

$$c_2(A) \quad -\frac{1}{8\pi^2}\int_M \mathrm{tr}(F_A \wedge F_A)$$

is related to linear invariant (first Chern class)

$$c_1(a) = +\frac{1}{2\pi}\int_{M/SO(3)} F_A \quad .$$

In fact, if A is sufficiently well behaved near infinity then

$$c_1(a) = c_2(A) \quad .$$

as can be verified by a direct calculation using (6.2.1).

Finally, if one defines a metric on $M/SO(3)$ by

$$ds^2 = R^{-2}(r,t)(g_r dt^2 + g_r dr^2) \tag{6.23}$$

then $\mathscr{A}_{M/SO(3)}(a,\Phi)$ is the action of the $U(1)$ abelian Higgs model on $M/SO(3)$ endowed with the Riemannian metric (6.23).

The Yang-Mills equations are satisfied by the connection A on M iff the abelian Higgs equations are satisfied by (a,Φ) on $M/SO(3)$. We leave the proof of these last statements as an exercise.

III. VORTICES

III.1. INTRODUCTION TO THE ABELIAN HIGGS MODEL

Consider the two-dimensional abelian Higgs (U(1)) Higgs model with action

$$\mathscr{A} = \frac{1}{2}\int_{\mathbb{R}^2}\left\{D_A\Phi \wedge *\overline{D_A\Phi} + F_A \wedge *F_A + \frac{\lambda}{4} *(\Phi\bar{\Phi} - 1)^2\right\} ,$$

where λ is a constant. In terms of components,

$$\mathscr{A} = \frac{1}{2}\int_{\mathbb{R}^2}\left\{|(\partial_j - iA_j)\Phi|^2 + \frac{1}{2} F_{jk}F_{jk} + \frac{\lambda}{4} (\Phi\bar{\Phi} - 1)^2\right\} d^2x. \quad (1.1)$$

The Higgs field $\Phi = \Phi_1 + i\Phi_2$ is a complex scalar field with Φ_1, Φ_2 real; alternatively it may be interpreted as a cross-section of a Hermitian line bundle over $\mathbb{R}^2$. The A_j are the components of the connection on the line bundle, and $F_{jk} = \partial_j A_k - \partial_k A_j$ is the curvature. Finite action requires that a configuration (A,Φ) have the following behavior:

$$|\Phi| \to 1 ,$$

$$D_A\Phi \equiv d\Phi - iA\Phi \to 0 , \qquad \text{as} \qquad |x| \to \infty . \quad (1.2)$$

As explained in Chapter II, if these limits are sufficiently uniform then the vortex number

$$N = \frac{1}{2\pi}\int_{\mathbb{R}^2} d^2x\, F_{12} = \frac{1}{2\pi}\int_{\mathbb{R}^2} F_A \quad (1.3)$$

is an integer. We prove applicability of these limits below.

The variational equations of the action $\mathscr{A}$ are

$$d*F_A = \frac{i}{2} * (\Phi\overline{D_A\Phi} - \overline{\Phi}D_A\Phi) \tag{1.4a}$$

$$D_A*D_A\Phi = \frac{\lambda}{2} * (\Phi\overline{\Phi} - 1)\Phi \tag{1.4b}$$

or equivalently, with $(\nabla_A)_k = \partial_k - iA_k$ and $\nabla_A^2 = \Sigma_{k=1}^2 (\nabla_A)_k^2$:

$$\partial_\mu F_{\mu\nu} = J_\nu \quad ,$$

$$\nabla_A^2\Phi = \frac{\lambda}{2}\Phi(|\Phi|^2 - 1) \quad ,$$

with

$$J_\nu = -\frac{i}{2}(\Phi(\partial_\nu + iA_\nu)\overline{\Phi} - \overline{\Phi}(\partial_\nu - iA_\nu)\Phi)$$

$$= \mathrm{Im}(\Phi(\overline{\nabla_A)_\nu\Phi}) \quad .$$

Bogomol'nyi pointed out that, when $\lambda = 1$, a lower bound on the action results from rewriting the action via an integration by parts as

$$\mathscr{A} = \int_{\mathbb{R}^2} d^2x \left\{ \frac{1}{2}[(\partial_1\Phi_1 + A_1\Phi_2) \mp (\partial_2\Phi_2 - A_2\Phi_1)]^2 \right.$$

$$+ \frac{1}{2}[(\partial_2\Phi_1 + A_2\Phi_2) \pm (\partial_1\Phi_2 - A_1\Phi_1)]^2$$

$$\left. + \frac{1}{2}\left[F_{12} \pm \frac{1}{2}(\Phi_1^2 + \Phi_2^2 - 1)\right]^2 \right\} \pm \frac{1}{2}\int_{\mathbb{R}^2} d^2x\, F_{12} \quad . \tag{1.5}$$

Then

$$\mathscr{A} \geq \pi|N| \quad . \tag{1.6}$$

If $N \geq 0$, then (1.6) is an equality if and only if

$$(\partial_1\Phi_1 + A_1\Phi_2) - (\partial_2\Phi_2 - A_2\Phi_1) = 0 \quad , \tag{1.7a}$$

$$(\partial_2\Phi_1 + A_2\Phi_2) + (\partial_1\Phi_2 - A_1\Phi_1) = 0 \quad , \tag{1.7b}$$

and

$$F_{12} + \frac{1}{2}(\Phi_1^2 + \Phi_2^2 - 1) = 0 \quad . \tag{1.7c}$$

If $N < 0$, then $\mathscr{A} = -\pi N$ if and only if

$$(\partial_1\Phi_1 + A_1\Phi_2) + (\partial_2\Phi_2 - A_2\Phi_1) = 0 \quad , \tag{1.8a}$$

$$(\partial_2\Phi_1 + A_2\Phi_2) - (\partial_1\Phi_2 - A_1\Phi_1) = 0 \quad , \tag{1.8b}$$

and

$$F_{12} - \frac{1}{2}(\Phi_1^2 + \Phi_2^2 - 1) = 0 \quad . \tag{1.8c}$$

Solutions to (1.7-8) are automatically solutions to the variational equations (1.4) with $\lambda = 1$.

Let

$$z = x_1 + ix_2, \qquad \partial_z = \frac{1}{2}(\partial_1 - i\partial_2), \qquad \partial_{\bar{z}} = \bar{\partial}_z = \frac{1}{2}(\partial_1 + i\partial_2) \ .$$

Furthermore, we let

$$\Phi = \Phi_1 + i\Phi_2 \quad , \qquad A = \alpha dz + \bar{\alpha} d\bar{z} \quad ,$$

where

$$\alpha = \frac{1}{2}(A_1 - iA_2) \ , \qquad \bar{\alpha} = \frac{1}{2}(A_1 + iA_2) \ .$$

In terms of the above definitions,

$$D_A\Phi = ((\partial_z - i\alpha)\Phi)dz + ((\partial_{\bar{z}} - i\bar{\alpha})\Phi)d\bar{z} \quad . \tag{1.9}$$

Then the equations (1.7a,b) become the real and imaginary parts of

$$D_A\Phi - i{*}D_A\Phi = 2((\partial_{\bar{z}} - i\bar{\alpha})\Phi)d\bar{z} = 0 \quad . \tag{1.10}$$

We remark that $*dz = -idz$. The first result is an existence theorem for equations (1.7a-c).

THEOREM 1.1. *Given an integer* $N \geq 0$ *and a set* $\{z_i\}$, $j = 1,2,\ldots,N$, *of* N *points in* $\mathbb{C}$, *there exists a finite action solution to equations* (1.7), *unique up to gauge equivalence, with the following properties:*

(i) *The solution is globally* C^∞.

(ii) *The zeros of* Φ *are the set of points* $\{z_j\}$.
$z \to z_j$

$$\Phi(z,\bar{z}) \sim c_j (z - z_j)^{n_j}, \qquad c_j \neq 0 , \tag{1.11}$$

where n_j *is the multiplicity of* z_j *in the set* $\{z_j\}$.

In addition for this solution,

(a) $|D_A\Phi| \leq \text{const}(1 - |\Phi|) \leq \text{const}\ \exp(-(1-\delta)|z|)$,

for any $0 < \delta$ *with* $\text{const} = \text{const}(\delta)$.

(b) $$N = \frac{1}{2\pi}\int_{\mathbb{R}^2} F_A = \sum_{\text{distinct } z_j} n_j = \pi^{-1}\mathscr{A} \quad . \tag{1.12}$$

Given integer $N < 0$, *the above statements hold for equations* (1.8) *but with* $\mathscr{A} = -N\pi$ *and with*

$$\bar{\Phi}(z,\bar{z}) \sim c_j (z - z_j)^{n_j} \qquad \textit{as} \qquad z \to z_j \ ,$$

and

$$N = -\sum_{\text{distinct } z_j} n_j = \frac{1}{2\pi}\int_{\mathbb{R}^2} F_A \quad .$$

Here $n_j > 0$, $c_j \neq 0$.

Remark. The solutions with $N > 0$ are called N-vortex solutions. Those with $N < 0$ are called N-antivortex solutions. We remark that antivortex solutions (A',Φ') are obtained from vortex solutions (A,Φ)

by the substitution $\alpha'(z) = -\bar{\alpha}(-\bar{z})$ and $\Phi'(z) = \Phi(-\bar{z})$. Thus we now restrict attention to $N > 0$. For $N = 0$, it follows from Theorem 1.1 that $\Phi = 1$, $A = 0$, modulo gauge transformations.

Theorem 1.1 is proved in §2-9. In §10 we pose the question: Do solutions exist to the variational equations (1.4) which are not solutions to (1.7) or (1.8)? The answer is no.

THEOREM 1.2 (Theorem 10.1). *Consider the action functional* $\mathscr{A}$ *of* (1.1) *with* $\lambda = 1$. *Every finite action critical point of* $\mathscr{A}$ *is a solution given in Theorem 1.1.*

Remark. Theorem 1.2 shows that mixed vortex-antivortex solutions do not exist.

As we showed in Chapter II, the assumption that the SU(2) Yang-Mills connection on $\mathbb{R}^4$ is O(3) - symmetric implies that it defines a finite action critical point of the abelian Higgs functional on the Poincaré half-plane, cf. Chapter II.6.

A straightforward generalization of Theorem 1.2 to this case is the statement that the only O(3) symmetric solutions to the SU(2) Yang-Mills equations on $\mathbb{R}^4$ are either self-dual or anti self-dual. This is the content of Theorem 11.1 of §11. Thus every O(3) symmetric solution to the SU(2) Yang-Mills equations is either N instantons or N anti-instantons on a line.

Question. Does a closed form representation for the vortex solutions (A,Φ) exist, such as the representations for instantons on a line? So far, one has not been found.

Let us return to the action functional $\mathscr{A}$ as a function of the parameter $\lambda > 0$. (The case $\lambda = 0$ is trivial by Corollary II.2.3c.) The critical points of $\mathscr{A}$ are expected to display three qualitatively different behaviors according to whether $\lambda <, =,$ or > 1. For $\lambda = 1$, we stated that the action of a solution is proportional to N. In other words, the N vortices can be thought of as each having a self-action π, with no interaction between distinct vortices.

For $\lambda \neq 1$, there is a heuristic picture of the critical points, explained in Chapter I.4, leading to repulsion of like vortices for $\lambda > 1$ and attraction of like vortices for $\lambda < 1$. This means that for $\lambda > 1$, a configuration with vortices is unstable to the vortices moving apart; a critical point of $\mathscr{A}$ will not be stable for $|N| > 1$. Likewise a configuration with N vortices and $\lambda < 1$ can be stationary only if the positions of all vortices coincide. Thus we are led to

CONJECTURE 1. *If* $\lambda < 1$, *there exists a finite action critical point* (A_N, Φ_N) *of the functional* $\mathscr{A}$ *of* (1.1) *for each integer* N, *with* $\Phi_N(0) = 0$. *This critical point is the only critical point of* $\mathscr{A}$, *modulo gauge equivalence and translations on* $\mathbb{R}^2$. *The point* (A_N, Φ_N) *is a minimum of* $\mathscr{A}$. *The field* (A_N, Φ_N) *is rotationally symmetric.*

Rotational symmetry means

$$\begin{aligned} A_N &= N\alpha(r)\,d\theta \quad , \\ \Phi_N &= \phi(r)e^{iN\theta} \quad , \\ \lim_{r\to\infty} \alpha(r),\ \phi(r) &= 1 \quad , \\ \lim_{r\to 0} \alpha(r),\ \phi(r) &= 0 \quad , \end{aligned} \tag{1.13}$$

where $\alpha(r), \phi(r) \in C^{\infty}(0,\infty)$ and (r,θ) are polar coordinates $\mathbb{R}^2$.

CONJECTURE 2. *If* $\lambda > 1$, $|N| > 1$, *the functional* $\mathscr{A}$ *has no finite action, stable critical points. It has an unstable critical point of the form* (1.13) *for each* N *with* $|N| > 1$, *which is the only critical point of* $\mathscr{A}$ *modulo gauge transformations and translations. For* $N = 0, \pm 1$, *the functional* $\mathscr{A}$ *has a stable critical point of the form* (1.13) *which is also unique in the same sense.*

Parts of Conjectures 1 and 2 have been established. One can prove that finite action critical points of $\mathscr{A}$ of the form (1.13) exist for every $N \in \mathbb{Z}$ and $\lambda > 0$. These critical points satisfy (1.2), (1.4).

The uniqueness of these critical points and the stability properties have not been rigorously established.

CONJECTURE 3. *Vortices of opposite charge (i.e., a vortex-antivortex pair) attracts for all* λ. *Hence for a time-independent solution they annihilate and are not observed. With angular momentum, however, a vortex-antivortex pair may form a time-dependent bound state (breather mode). It would be interesting to find such solutions.*

THEOREM 1.3. (Theorem V.1.1) *Let* (A,Φ) *be a finite action solution to* (1.4) *with* $\lambda > 0$. *Assume that there exists a gauge in which the components of* (A,Φ) *and their first derivatives are locally square integrable. Then* (A,Φ) *is gauge equivalent to a globally* C^∞ *solution and locally gauge equivalent to a real analytic solution (in the variables* (x_1,x_2)*).*

Next we come to the question of how solutions to (1.4) decay as $|x| \to \infty$. It is here that we begin to justify the formal picture of the Higgs effect in Chapter I.3. We find decay rates m_L, m_T for the longitudinal and transverse parts of $D_A\Phi$, $\mathrm{Re}(\bar{\Phi}D_A\Phi)$ and $\mathrm{Im}(\bar{\Phi}D_A\Phi)$, respectively. These also apply to $1-|\Phi|^2$ and F_A.

THEOREM 1.4. (cf. Theorem 9.1) *Let* (A,Φ) *be a smooth finite action solution to the second order equations* (1.4). *Then given* $\varepsilon > 0$, *there exists* $M = M(\varepsilon,(A,\Phi)) < \infty$ *such that*

$$|\mathrm{Re}(\bar{\Phi}D_A\Phi)| + 1 - |\Phi|^2 \le Me^{-(1-\varepsilon)m_L|x|} \tag{1.14}$$

and

$$|\mathrm{Im}(\bar{\Phi}D_A\Phi)| + |F_A| \le Me^{-(1-\varepsilon)m_T|x|} . \tag{1.15}$$

Here $m_L = \min(\lambda^{\frac{1}{2}},2)$, $m_T \equiv 1$.

<u>Remark</u>. Since $|\Phi| \to 1$ by (1.14), the decay rates for $(\bar{\Phi}D_A\Phi)$ are decay rates for the longitudinal and transverse parts of $D_A\Phi$. Furthermore, the inequality

$$m_L \le 2m_T$$

expresses the instability of the Higgs particle mass m_L against decay into 2 photons of mass m_T. In superconductivity theory m_L^{-1} is the correlation length and m_T^{-1} is penetration depth. The ratio m_L/m_T is the important quantity, and in our units m_T is scaled to one. It is interesting to note that this classical picture extends to statical mechanics (and hence also to quantum field theory) cf. [2].

As a corollary to exponential decay, we have by the general results of Chapter II,

COROLLARY 1.5. *Let* (A,Φ) *be a smooth finite action solution to* (1.4). *Then*

$$N \equiv \frac{1}{2\pi}\int_{\mathbb{R}^2} F$$

is an integer.

In Chapter I, we described how in the theory of superconductivity the Higgs field Φ plays the role of an order parameter. The completely ordered state has $|\Phi| = 1$ and $|F| = 0$; this state is superconducting and has expelled the magnetic field, namely F. The completely disordered state has $|\Phi| = 0$ and $|F|$ large, such as occurs in the center of a vortex. We now give a mathematical bound which reflects the role of Φ as an order parameter in the classical equations.

THEOREM 1.6. (Theorem 8.1) *Let* (A,Φ) *be a smooth, finite action solution to the second order equations* (1.4a,b) *with* $0 < \lambda$. *Then*

$$|\Phi| < 1 \tag{1.16}$$

and if $\lambda \le 1$,

$$|F| + |D\Phi| \le 4(1 - |\Phi|) \quad . \tag{1.17}$$

III.2. THE FIRST ORDER EQUATIONS

The first order equations (1.7) for (A,Φ) can be reduced to a single nonlinear elliptic equation for the unknown $\ln|\Phi|^2$. This reduction serves two purposes. First the equation obtained in this manner can be handled by variational techniques. The second useful feature is that the poles of $\ln|\Phi|^2$ are the zeros of Φ. This fact allows us to find solutions with specified zeros, and to characterize every solution by its zeros.

In this section we deal with (A,Φ) which are smooth, an assumption we justify later, cf. Theorem 1.3.

In order to derive our new equations write (1.10) as

$$\alpha = i\partial_z \ln \bar{\Phi} \quad , \quad \Phi \neq 0 . \tag{2.1}$$

Because α is smooth, it extends by continuity to the zero set of Φ. Define a real, single valued function u and, a real, multivalued function Θ by

$$\Phi = \exp \frac{1}{2}\left[(u + i\Theta)\right] , \tag{2.2}$$

or equivalently $u + i\Theta = 2 \ln \Phi$. Then

$$\alpha = \frac{i}{2} \partial_z (u - i\Theta) . \tag{2.3}$$

At this point we consider the function Θ, which can also be written

$$\frac{\Phi}{|\Phi|} = e^{i\Theta/2} . \tag{2.4}$$

Following the discussion in Chapter II.3, Θ is not uniquely determined by the equations. Rather, it is determined modulo a C^∞ function, i.e. modulo a smooth gauge transformation. If the limits of (A,Φ) are achieved sufficiently rapidly as $|z|\to\infty$, we also showed in Chapter II.3 that Φ defines a map ϕ of the circle (at ∞) to the circle, namely for $\theta \equiv \arg z$,

$$\phi(\theta) = \lim_{|z|\to\infty} \Phi(|z|,\theta)$$

$$= \lim_{|z|\to\infty} e^{\frac{1}{2} i\Theta(|z|,\theta)} \quad . \tag{2.5}$$

The winding number N of the map ϕ equals the vortex number

$$N = \frac{1}{2\pi}\int F = \text{winding number of } \phi \ . \tag{2.6}$$

Thus for $N \neq 0$, it follows by homotopy considerations that Φ must have a zero and that $\Theta(|z|,\theta)$ cannot be smooth. For $|z|$ sufficiently large we can require Θ to be smooth and to satisfy

$$\frac{1}{2}\Theta(|z|,\theta+2\pi) = \frac{1}{2}\Theta(|z|,\theta) + 2\pi N \quad . \tag{2.7}$$

In order to proceed, we characterize the singular points of Θ and the zeros of Φ.

DEFINITION 2.1. *Let* (A,Φ) *be a smooth solution to the first order equations* (1.7). *Define the zero set* $Z(\Phi)$ *and the singular set* $S(\Theta)$ *by*

$$Z(\Phi) = \{z \in \mathbb{C}:\ \Phi(z) = 0\} \quad , \tag{2.8}$$

$$S(\Theta) = \mathbb{C} \setminus \{z \in \mathbb{C}:\ \Theta(z) \in C^{\infty}\} \quad . \tag{2.9}$$

Note that while Θ is gauge dependent, both $Z(\Phi)$ and $S(\Theta)$ are invariant under smooth gauge transformations. We can characterize Z and S for any solution:

THEOREM 2.2. *Let* (A,Φ) *be a finite action solution to the first order equations* (1.7) *in which the components of* (A,Φ) *are locally square integrable. Then* (A,Φ) *is smooth and there exist* N *(not necessarily distinct) points in* $\mathbb{C}$, $\{z_1,\ldots,z_N\}$, *such that*

$$Z(\Phi) = S(\Theta) = \{z_1,\ldots,z_N\} \quad . \tag{2.10}$$

There is a neighborhood of each z_k *in which*

$$\Phi(z) = (z-z_k)^{n_k} h_k(z) \quad , \tag{2.11}$$

where n_k *is the multiplicity of the point* z_k *in* $\{z_1,\ldots,z_N\}$, *and* $h_k(z)$ *is a* C^∞, *nonvanishing function.*

The proof of Theorem 2.2 is given in §5, and relies on what is known in complex analysis as the $\bar{\partial}$-Poincaré lemma.

Fix the points $\{z_1,\ldots,z_N\}$; we find a smooth solution (A,Φ) with zeros $Z(\Phi) = \{z_1,\ldots,z_N\}$, cf. Theorem 2.3 below. This solution is *unique* up to smooth gauge transformations and satisfies $N = (2\pi)^{-1}\int F$. Because the solutions of (1.7) are characterized by $Z(\Phi)$, cf. Theorem 2.2, we have thus found all solutions to (1.7). By (2.11) with $c_k = h_k(z_k) \neq 0$,

$$\Phi \sim c_k(z-z_k)^{n_k} \tag{2.12}$$

as $z \to Z(\Phi)$. Here n_k is the multiplicity of z_k in $Z(\Phi)$. By a smooth gauge transformation, we can take c_k to be real.

Motivated by Theorem 2.2, we prepare to state Theorem 2.3. We now regard (2.2) as defining Φ in terms of u and Θ. Set

$$\Theta \equiv 2\sum_{k=1}^{N} \arg(z-z_k) \quad , \tag{2.13}$$

and

$$u_0 \equiv -\sum_{k=1}^{N} \ln(1+\mu|z-z_k|^{-2}) \ , \tag{2.14}$$

where $\mu > 4N$. The function u_0 is C^∞ except at $\{z_1,\dots,z_N\}$.

With Θ given by (2.13), and u defined by (2.2), equation (1.7c) becomes

$$-\Delta u + e^u - 1 = -4\pi \sum_{k=1}^{N} \delta(z - z_k) \quad , \tag{2.15}$$

with the boundary condition

$$\lim_{|z|\to\infty} u = 0 \quad , \tag{2.16}$$

resulting from $|\Phi| \to 1$. (In fact we show in §8 that $\mathscr{A}(A,\Phi) < \infty$ is sufficient to ensure exponential convergences of $|\Phi(x)| \to 1$.) In (2.15), δ is the Dirac measure. As a distribution, u_0 satisfies

$$-\Delta u_0 = 4 \sum_{k=1}^{N} \frac{\mu}{(|z-z_k|^2+\mu)^2} - 4\pi \sum_{k=1}^{N} \delta(z - z_k) \; . \tag{2.17}$$

Define $g_0 \in C^\infty(\mathbb{R}^2)$ by

$$g_0 = 4 \sum_{k=1}^{N} \frac{\mu}{(|z-z_k|^2+\mu)^2} \; . \tag{2.18}$$

In terms of

$$v \equiv u - u_0 \quad , \tag{2.19}$$

a solution to (2.15-6) is equivalent to a solution v to the equation

$$\Delta v = e^{v+u_0} + (g_0 - 1) \tag{2.20}$$

with

$$\lim_{|z|\to\infty} v = 0 \; . \tag{2.21}$$

If $v \in C^\infty(\mathbb{R}^2)$, then the poles of u are those of u_0, and with Φ defined by (2.2), $Z(\Phi) = \{z_1,\dots,z_N\}$.

THEOREM 2.3. (cf. §4) *Let* $\{z_1,\ldots,z_N\}$ *be a point in the* N*-fold symmetric product of* $\mathbb{C}$, *and let* $\mu > 4N$. *Then there exists a unique function* $v(z)$ *which satisfies* (2.20-21). *The sum* $u = u_0 + v$ *is independent of* μ. *Furthermore:*

a) *The function* v *is real analytic in* (x_1,x_2), *when we write* $z = x_1 + ix_2$.

b) *The functions* (α,Φ) *defined by* (2.1-2) *give a real analytic solution (in* x_1,x_2*) to the first order equations* (1.7).

c) $N = \frac{1}{2\pi}\int F$.

While we postpone most aspects of the proof, we insert a few comments here. In particular the analyticity of v ensures (by the form of u_0) that

$$u - \sum_{k=1}^{N} \ln|z - z_k|^2 \tag{2.22}$$

is real analytic in (x_1,x_2). From this we infer the regularity of (α,Φ) as follows: For $z \notin Z(\Phi)$, the functions Θ and β, with β defined by

$$\beta \equiv \sum_{k=1}^{N} \ln|z - z_k|^2 \quad , \tag{2.23}$$

are conjugate harmonic functions. Thus

$$\partial_z(\beta - i\Theta) = 0 \quad , \qquad z \notin Z(\Phi) \ . \tag{2.24}$$

It follows that α defined by (2.1) is real analytic in (x_1,x_2). Furthermore $\exp[(\beta + i\Theta)/2]$ is analytic in z and $\exp[(u - \beta)/2]$ is real analytic in (x_1,x_2), so Φ is real analytic in (x_1,x_2).

To recover the vortex number N, we appeal to Theorem 1.4 which ensures good behavior of (A,Φ) at infinity. Hence by Chapter II.3,

$$\text{vortex number} = \frac{1}{2\pi}\int F = \text{winding number } \phi.$$

Given Θ in (2.13), we read off that the winding number equals N.

III.3. $-\Delta u + e^{u} - 1 = -4\pi \sum_j \delta(x - x_j)$

In this section we establish *a priori* estimates for this equation on $\mathbb{R}^2$. We return here to the standard Cartesian coordinates (x_1, x_2) on $\mathbb{R}^2$ and study u in terms of the equation (2.20) for v. This equation is the variational equation for the functional

$$a(v) = \int_{\mathbb{R}^2} \left[\frac{1}{2}|\nabla v|^2 + v(g_0 - 1) + e^{u_0}(e^{v} - 1)\right] dx$$

$$= \frac{1}{2}\|\nabla v\|^2_{L_2} + \langle g_0 + e^{u_0} - 1, v\rangle_{L_2} + \langle e^{u_0}, e^{v} - 1 - v\rangle_{L_2} . \qquad (3.1)$$

We work in this section with v an element of the real Sobolev space H_1.

DEFINITION 3.1. *The Sobolev space* H_1 *of real functions is the Hilbert space of real functions defined by the inner product*

$$\langle f, g\rangle_{H_1} = \langle f, g\rangle_{L_2} + \langle \nabla f, \nabla g\rangle_{L_2} , \qquad (3.2)$$

and norm

$$\|f\|_{H_1} = \langle f, f\rangle_{H_1}^{1/2} . \qquad (3.3)$$

Let us start by justifying the choice of the space H_1. By appealing to the regularity estimates of Chapter V, we restrict attention to smooth, finite action solutions (A, Φ).

THEOREM 3.2. *Let* (A, Φ) *be a smooth, finite action solution to* (1.7) *with zeros* $Z(\Phi) = \{z_1, \ldots, z_N\}$. *Let* u_0 *be defined by* (2.14) *and* $v \equiv \ln|\Phi|^2 - u_0$. *Then*

$$v \in H_1 .$$

Proof. Since Φ is smooth, $\ln|\Phi|^2$ is smooth except at $Z(\Phi)$. In a neighborhood of $Z(\Phi)$, we use the representation (2.11) of Theorem 2.2. By the explicit form of u_0, and the fact that $h_k(z) \neq 0$, we see that $v \in C^\infty$. Thus to prove $v \in H_1$, we need only prove bounds on v and ∇v which ensure their square integrability. For large $|x|$, the function u_0 is $O(|x|^{-2})$, and its derivative $O(|x|^{-3})$; both are L_2. By the decay estimates of Theorem 1.4, $0 \le 1 - |\Phi|^2 \le O(1)\exp(-(1-\varepsilon)|x|)$, so for large $|x|$,

$$\ln|\Phi|^2 \le O(1)\exp(-(1-\varepsilon)|x|) \in L_2 \quad ,$$

from which we infer $v \in L_2$. Finally, we estimate ∇v using the identity

$$\nabla v = e^{-u}(\nabla|\Phi|^2) - \nabla u_0 \quad .$$

Since $u \to 0$ uniformly, by (1.14), and since ∇u_0 is L_2 outside a sufficiently large ball, we need only show that $\nabla|\Phi|^2$ is L_2 outside a large ball, but with the notation of Chapter VI.5,

$$\begin{aligned} \nabla|\Phi|^2 &= (\bar\Phi\nabla\Phi + \Phi\nabla\bar\Phi) \\ &= (\bar\Phi\nabla_A\Phi) + \Phi(\overline{\nabla_A\Phi}) = 2\,\mathrm{Re}(\bar\Phi\nabla_A\Phi) \quad , \end{aligned}$$

which by (1.1) is exponentially small. This completes the proof.

Having shown that it is sufficient to study the functional (3.1) for $v \in H_1$, we next prove that $a(v)$ is defined for all $v \in H_1$.

PROPOSITION 3.3. *The functional* $a(v)$ *is defined for* $v \in H_1$. *For some* $c < \infty$,

$$|a(v)| \le \exp(c\|v\|_{H_1}^2) \quad . \tag{3.4}$$

Proof. It is sufficient to establish (3.4) for all $v \in C_0^\infty$, with a constant c independent of v. It then extends by continuity to all of H_1.

LEMMA 3.4. *With* g_0 *of* (2.18) *and* u_0 *of* (2.14),

$$u_0,\ g_0,\ e^{u_0} - 1 \tag{3.5}$$

are all L_2. *Also* $e^{u_0} \le 1$.

Proof. To bound g_0, note by (2.18) that it is $O((1+|x|^4)^{-1})$, and hence L_2. Now return to u_0. By (2.14), $u_0 \le 0$, so $e^{u_0} \le 1$. Consider $R > \max|z_k| = \max|x_k|$. For $|x| \ge R$, we have by (2.14),

$$|u_0| + |e^{u_0} - 1| \le O(|x|^{-2}) \in L_2 \quad .$$

For $|x| \le R$, the function $0 \le 1 - e^{u_0} \le 1$, so it is L_2. The function u_0 has logarithmic singularities $x = x_k$, and is bounded on a complement of a neighborhood of the x_k's. Since a logarithmic singularity is locally square integrable, so is u_0. This completes the proof.

Let us now return to the proof of (3.4). We show that each of the three terms in (3.1) satisfies a bound of the form (3.4). For the first term there is nothing to prove, since $\|\nabla v\|_{L_2} \le \|v\|_{H_1}$. For the second term,

$$|\langle v, g_0 + e^{u_0} - 1\rangle_{L_2}| \le \|v\|_{L_2} (\|g_0\|_{L_2} + \|e^{u_0} - 1\|_{L_2})$$

$$\le \text{const } \|v\|_{H_1} \quad ,$$

using the lemma. To bound the third term, again we use the lemma.

$$|\langle e^{u_0}, e^v - 1 - v\rangle_{L_2}| \le \|e^{u_0}\|_{L_\infty} \|e^v - 1 - v\|_{L_1}$$

$$\le \|e^v - 1 - v\|_{L_1} \le \exp(c\|v\|^2_{H_1}) \ . \tag{3.6}$$

For the last inequality in (3.6), we appeal to Chapter VI, Proposition 2.3, and the proof is complete.

Next we consider the directional derivative $(D_h a)(v)$ of the functional $a(v)$ in direction h. For $v, h \in H_1$,

$$(D_h a)(v) = \lim_{t\to 0} \frac{a(v+th) - a(v)}{t} .$$

See Chapter VI.7 for properties of functionals and derivatives.

PROPOSITION 3.5. *The functional* $a(v)$ *is differentiable on* H_1. *For some* $c < \infty$, $|(D_h a)(v)| \le \|h\|_{H_1} \exp(c\|v\|^2_{H_1})$, *and*

$$|a(v+h)-a(v)-(D_h a)(v)| \le O(1) \|h\|^2_{H_1} \exp(c\|v\|^2_{H_1} + \|h\|^2_{H_1}) . \tag{3.7}$$

Furthermore,

$$(D_h a)(v) = \langle\nabla v,\nabla h\rangle_{L_2} + \langle g_0 + e^{u_0} - 1, h\rangle_{L_2} + \langle e^{u_0}(e^v - 1), h\rangle_{L_2} . \tag{3.8}$$

Proof. Assuming the derivative of $a(v)$ exists, it is formally given by (3.8). To establish this identity, note that each of the three terms in (3.1) can be differentiated individually in H_1. The linear term yields the second term in (3.8). For the quadratic term,

$$t^{-1}\frac{1}{2}(\|\nabla(v+th)\|^2_{L_2} - \|\nabla v\|^2_{L_2}) - \langle\nabla v,\nabla h\rangle_{L_2}$$

$$= \frac{1}{2} t \|\nabla h\|^2_{L_2} \le \frac{1}{2} t \|h\|^2_{H_1} \to 0 . \tag{3.9}$$

For the third term, note that

$$t^{-1}[e^{v+th} - e^v - th] = e^v(e^{th} - 1)t^{-1} - h$$

$$= (e^v - 1)h + \sum_{n=2}^{\infty} e^v \frac{t^{n-1}h^n}{n!} \tag{3.10}$$

Insertion of the second term, i.e. the error term in (3.10), in the difference quotient of (3.1), and using $u_0 \le 0$ and the Schwarz inequality we infer

$$\left|\sum_{n=2}^{\infty}\frac{t^{n-1}}{n!}\int e^{u_0+v}h^n\,dx\right| \le t\sum_{n=2}^{\infty}\frac{t^{n-2}}{n!}\left(\int h^n\,dx+\left(\int h^{2n}\,dx\right)^{\frac{1}{2}}\sum_{m=1}^{\infty}\frac{1}{m!}\left(\int v^{2m}\,dx\right)^{\frac{1}{2}}\right)$$

$$\le t\ \mathrm{const}\|h\|_{H_1}^2\left[\exp\ \mathrm{const}\left(\|h\|_{H_1}^2 t^2+\|v\|_{H_1}^2\right)\right] \qquad (3.11)$$

In the last step we use the estimate of Proposition VI.2.3. As $t\to 0$, the error (3.11) vanishes, completing the proof of existence of the $(D_h a)(v)$ limit.

The proof of continuity of $(D_h a)(v)$ and linearity in h follow by similar estimates on

$$(D_h a)(v)-(D_{h'}a)(v') = (D_{h-h'}a)(v)+(D_{h'}a)(v)-(D_{h'}a)(v').$$

Expanding in series yields the bound on $|(D_h a)(v)|$. Isolating the terms linear in $h-h'$ or $v-v'$ yields the proof of Lipshitz continuity. The upper bounds are also similar; the $O(h^2)$ behavior in (3.7) is apparent by setting $t=1$ in (3.11).

We now study the "radial" derivative of $a(v)$, defined to be

$$(D_v a)(v) = \langle\nabla v,\nabla v\rangle + \langle v, e^{u_0+v}+g_0-1\rangle_{L_2}, \qquad (3.12)$$

and sometimes written $a'(v,v)$. In fact, for $\|v\|_{H_1}$ sufficiently large, (3.12) is strictly positive. This follows from a lower bound on $(D_v a)(v)$.

PROPOSITION 3.6. *There exist constants* $\varepsilon>0$ *and* $c<\infty$ *such that for all* $v\in H_1$,

$$-c+\varepsilon\|v\|_{H_1} \le (D_v a)(v) . \qquad (3.13)$$

LEMMA 3.7. *For all* x,

$$0 \le 1 - g_0(x) - e^{u_0(x)} \tag{3.14}$$

and there exists a constant $b \in (0,1)$ *such that*

$$b \le 1 - g_0(x) \quad . \tag{3.15}$$

Proof. Note

$$0 < g_0(x) \le 4N\mu^{-1} < 1 \tag{3.16}$$

so (3.15) follows. Define

$$\mu_k = \frac{\mu}{\mu + |z - z_k|^2} < 1 \quad . \tag{3.17}$$

Using $\mu_k < 1$, $4N\mu^{-1} < 1$, it follows that

$$g_0 + e^{u_0} = \prod_{k=1}^{N} (1-\mu_k) + \left(\frac{4N}{\mu}\right) \frac{1}{N} \sum_{k=1}^{N} \mu_k^2$$

$$\le \prod_k (1-\mu_k) + \frac{1}{N} \sum \mu_k^2 \quad .$$

For $0 \le x_k \le 1$,

$$\prod_{k=1}^{N} x_k \le \bar{x} \equiv \frac{1}{N} \sum_{k=1}^{N} x_k \quad . \tag{3.18}$$

Thus

$$g_0 + e^{u_0} \le 1 - \frac{1}{N} \sum_{k=1}^{N} \mu_k (1 - \mu_k) \le 1 \quad , \tag{3.19}$$

to complete the proof of (3.14).

LEMMA 3.8. *There exist constants* $b \in (0,1)$, $c < \infty$, *such that*

$$(D_v a)(v) \geq \|\nabla v\|_{L_2}^2 + b \int_{R^2} \frac{v^2}{1+|v|}\, dx - c \quad . \tag{3.20}$$

<u>Proof.</u> Consider the difference

$$M(v) \equiv (D_v a)(v) - \|\nabla v\|_{L_2}^2 = \int_{R^2} v(e^{u_0+v} + g_0 - 1)\, dx$$

$$= M(v_+) + M(-v_-) \quad , \tag{3.21}$$

with $v = v_+ - v_-$ the decomposition of v into its positive and negative parts. To bound $M(v_+)$, use the pointwise inequality

$$e^x - 1 - x \geq 0 \quad , \qquad x \in R \quad . \tag{3.22}$$

Thus

$$e^{u_0+v_+} + g_0 - 1 = \left[e^{u_0+v_+} - 1 - (u_0+v_+)\right] + v_+ + u_0 + g_0$$

$$\geq v_+ + u_0 + g_0 \quad ,$$

and

$$M(v_+) \geq \int v_+^2\, dx + \int v_+(u_0 + g_0)\, dx$$

$$\geq \frac{1}{2} \int v_+^2\, dx - \frac{1}{2} \int (u_0 + g_0)^2\, dx \quad .$$

But u_0, g_0 are L_2 functions, cf. (3.5), so

$$M(v_+) \geq \frac{1}{2} \int v_+^2\, dx - c \geq \frac{1}{2} \int \frac{v_+^2}{1+v_+}\, dx - c \quad . \tag{3.23}$$

Next consider $M(-v_-)$ and use the bound

$$1-e^{-x} \geq \frac{x}{1+x} \quad , \qquad x \geq 0 \quad . \tag{3.24}$$

Then

$$\begin{aligned} M(-v_-) &= \langle v_-, 1-g_0-e^{u_0}\rangle + \langle v_-, e^{u_0}(1-e^{-v_-})\rangle \\ &\geq \langle v_-, \left\{1-g_0-e^{u_0}+v_-(1+v_-)^{-1}e^{u_0}\right\}\rangle \\ &\geq b\int \frac{v_-^2}{1+v_-}\,dx \quad . \end{aligned} \tag{3.25}$$

The last inequality follows from using Lemma 3.7 on the expression in brackets:

$$\begin{aligned} (1+v_-)(1-g_0-e^{u_0}) + v_-e^{u_0} &= (1-g_0)v_- + (1-g_0-e^{u_0}) \\ &\geq (1-g_0)v_- \geq bv_- \quad . \end{aligned}$$

Adding (3.23) and (3.25) establishes the lemma.

Proof of Proposition 3.6. Using the Schwarz inequality, and Proposition VI.2.4,

$$\begin{aligned} (\int v^2\,dx)^2 &\leq \int \frac{v^2}{1+|v|}\,dx \int (v^2+|v|^3)\,dx \\ &\leq 2\int \frac{v^2}{1+|v|}\,dx \,\|v\|_{H_1}^2\,(1+\|v\|_{H_1}) \quad . \end{aligned} \tag{3.26}$$

Define $\sigma = \sigma(v)$ by

$$\int v^2 = \sigma \|v\|_{H_1}^2 \quad .$$

Then $0 < \sigma < 1$, and

$$\|\nabla v\|^2_{L_2} = (1-\sigma)\|v\|^2_{H_1} \quad .$$

Thus by Lemma 3.8 and (3.26),

$$(D_v a)(v) \geq (1-\sigma)\|v\|^2_{H_1} + \left(\frac{b}{2}\right) \frac{\sigma^2\|v\|^2_{H_1}}{(1+\|v\|_{H_1})} - c \quad , \tag{3.27}$$

with σ chosen appropriately. Note $b<1$. Then, as a function of $\sigma \in [0,1]$, the minimum of (3.27) occurs for $\sigma = 1$. Therefore

$$(D_v a)(v) \geq \left(\frac{b}{2}\right) \frac{\|v\|^2_{H_1}}{(1+\|v\|_{H_1})} - c \quad . \tag{3.28}$$

Choosing new constants b, c yields (3.13).

III.4. EXISTENCE OF VORTEX SOLUTIONS

We use the estimates of §3 to establish existence and uniqueness of solutions to the equation $-\Delta u + e^u - 1 = -4\pi \Sigma \delta_{x_i}$ which characterizes vortices. We refer to analytical tools in Chapter VI, §7-8.

PROPOSITION 4.1. *The functional* $a(v)$ *is a strictly convex functional of* v.

Proof. For $v, w \in H_1$, $t \in (0,1)$, we prove

$$a(tv+(1-t)w) < ta(v) + (1-t)a(w) \quad . \tag{4.1}$$

Since a sum of a strictly convex function and a linear function is strictly convex, we treat separately the three terms in (3.1). A quadratic functional is strictly convex by Proposition VI.7.9. Hence we need only treat the third term in (3.1). Pointwise, the function $e^v - 1 - v$ has a positive second derivative, and hence it is convex. Since $\exp u_0$ is positive, the third term in (3.1) is convex, completing the proof.

PROPOSITION 4.2. *The functional* $a(v)$ *is weakly lower semicontinuous on* H_1.

Proof. This is a consequence of convexity and the existence of the derivative; see Proposition VI.7.8.

A strictly convex, weakly lower semicontinuous functional $a(v)$ which satisfies the coercivity estimate (3.13) has a unique minimizing function $\bar{v}$. Furthermore, $\bar{v}$ is the only critical point of $a(v)$. (See Propositions VI.7.7-8.) Thus we have established

THEOREM 4.3. *The functional* $a(v)$ *has a single critical point* $\bar{v}$, *which satisfies*

$$a(\bar{v}) = \inf_{v \in H_1} a(v) \quad . \tag{4.2}$$

COROLLARY 4.4. *The function* $\bar{v}$ *satisfies* (2.20). *The function* $u = u_0 + \bar{v}$ *satisfies* (2.15). *The function* $\bar{v}$ *is independent of* μ *in* (2.14).

Proof. It follows from (4.2) and Proposition VI.7.5 that $(D_h a)(\bar{v}) = 0$ for all h. Hence by (3.8), the equation (2.20) holds in the weak sense, i.e. with $\Delta\bar{v}$ interpreted as a distribution derivative. Since (2.20) and (2.15) are related by translation, $u = u_0 + \bar{v}$ satisfies (2.15). Since every H_1 solution to (2.20) is a critical point of $a(v)$, and since $a(v)$ has only one critical point, the equation (2.20) has only one H_1 solution. Finally if $u_0(\mu_i)$, $i = 1,2$ are given, note $u_0(\mu_1) - u_0(\mu_2) \in H_1$. Hence $\bar{v}_1 + u_0(\mu_1) - \mu_0(\mu_2)$ satisfies (3.1) with $u_0(\mu_2)$. By uniqueness it equals $\bar{v}_2$. Thus $u_0 + \bar{v}$ is independent of μ.

THEOREM 4.5. *The solution* $\bar{v} \in H_1$ *is real analytic on* $\mathbb{R}^2$.

<u>Proof</u>. Since $\bar{v}$ is a weak solution to (2.20), $\Delta\bar{v} \in L_2$. Integration by parts gives $v \in H_2$ (Definition VI.1.1). Thus $\bar{v}$ is continuous, cf. Propositions VI.1.6 and VI.2.4. Repeating this argument for derivatives of $\bar{v}$, we obtain all pointwise derivatives of $\bar{v}$. Since solutions to elliptic equations are real analytic (see Morrey), so is $\bar{v}$. This completes the proof of Theorems 4.5 and 2.3.

III.5. ZEROS OF Φ

The interpretation of the zeros of the Higgs field Φ as the location of vortices arises from a representation of Φ stated in (2.11). The same representation is crucial for classifying all solutions to the first order equations for (A,Φ). In this section we prove this basic representation and its consequences, thus establishing Theorem 2.2.

PROPOSITION 5.1. *Let* (α,Φ) *be a smooth solution to the first order equation*

$$(\bar{\partial}_z - i\bar{\alpha})\Phi = 0 \quad . \tag{5.1}$$

Then the set of zeros $Z(\Phi)$ *is discrete and in some neighborhood of each* $z_k \in Z(\Phi)$,

$$\Phi(z) = (z - z_k)^{n_k} h_k(z) \quad , \tag{2.11),(5.2}$$

where n_k *is the (integer) order of the zero and* $h_k(z)$ *is* C^∞ *and nonvanishing on the neighborhood.*

LEMMA ($\bar{\partial}$-Poincaré lemma). *Let* $i\alpha(z)$ *be a* C^∞ *function on a closed disc* $B \subset \mathbb{C}$. *Then the differential equation*

$$\bar{\partial}_z \omega(z) = i\alpha(z) \tag{5.3}$$

has a C^∞ *solution* $\omega(z)$ *in the interior of* B. *In fact*

$$\omega(z) = \frac{1}{2\pi} \int_B \frac{\alpha(\zeta)}{\zeta - z}\, d\zeta \wedge d\bar{\zeta} \quad .$$

Proof. See, e.g. Griffiths and Harris [3].

Proof of Proposition 5.1. Let ω solve (5.3). Then

$$\Omega(z) = e^{-\omega(z)}\Phi(z)$$

is complex analytic, since

$$\bar{\partial}_z(e^{-\omega}\Phi) = e^{-\omega}(\bar{\partial}_z\Phi - (\bar{\partial}_z\omega)\Phi) = e^{-\omega}(\bar{\partial}_z - i\bar{\alpha})\Phi = 0 .$$

Thus

$$\Phi = e^{\omega}\Omega$$

is the product of a C^∞ nonvanishing function e^{ω} and a complex analytic function Ω. Since the zeros of a complex analytic function are discrete, a finite number of zeros occur in any bounded set B. Let z_k be such a zero of degree n_k. Then in a neighborhood of z_k, there is a nonvanishing analytic function Ω_k such that

$$\Omega(z) = (z - z_k)^{n_k}\Omega_k(z) .$$

Let $h_k = \Omega_k e^{\omega}$, so on this neighborhod

$$\Phi(z) = h_k(z)(z - z_k)^{n_k} , \quad h_k \neq 0 ,$$

as claimed.

PROPOSITION 5.2. *Let* (A,Φ) *be a smooth, finite action solution to the first order equations* (1.7). *Then* Φ *has a finite number of zeros* $Z(\Phi) = \{z_1,\dots,z_N\}$.

Proof. For a finite action smooth solution, we appeal to the decay (1.14) of Theorem 1.4. Thus $0 \le 1 - |\Phi| \le O(\exp(-|z|/2))$, so Φ is nonzero outside some disc B. By Proposition 5.1, only a finite number of zeros occur on B, and the proof is complete.

PROPOSITION 5.3. *Let* (A,Φ) *be a smooth, finite action solution to* (1.7). *Then there exist* N *points* $\{z_1,\dots,z_N\}$ *such that*

$$Z(\Phi) = S(\Theta) = \{z_1,\dots,z_N\} .$$

Remark. By Theorem 2.3, N is the vortex number of (A,Φ).

Proof. Since Φ is smooth, any singularity of Θ must be a zero of Φ. Hence $S(\Theta) \subset Z(\Phi)$. Conversely, suppose $z_k \in Z(\Phi)$. By Proposition 5.1, the representation (5.2) holds near z_k, so

$$\begin{aligned}\Phi(z) &= |z-z_k|^{n_k} \, |h_k(z)| e^{in_k \arg(z-z_k) + i \arg h_k(z)} \\ &= e^{(u+i\Theta)/2} .\end{aligned}$$

Thus in a neighborhood of z_k, $\Theta = 2n_k \arg(z-z_k) + 2 \arg h_k(z)$, mod 2π. But h_k does not vanish and $h_k \in C^\infty$, so $\arg h_k \in C^\infty$. Hence $z_k \in S(\Theta)$. Thus the proof of $Z = S$ is complete, as is that of Theorem 2.2.

III.6 BASIC IDENTITIES

We start from the second order equations (1.4) and derive some identities for the solutions. These identities are useful in establishing properties of the solutions. Here $d = 2$ and p denotes the degree of a form.

To simplify notation, define f, w, g by

$$f = *F , \qquad w = \frac{1}{2}(1 - |\Phi|^2) , \qquad g = D_A\Phi . \tag{6.1}$$

The second order equations (1.4) can be written

$$*df = \operatorname{Im}(\Phi\bar{g}) = -\frac{i}{2}(\Phi\bar{g} - \bar{\Phi}g) \qquad (1.4a),(6.2)$$

and

$$-*D_A*D_A\Phi = \lambda\Phi w . \qquad (1.4b),(6.3)$$

Note $** = (-1)^{p(d-p)}$. See Chapter VI.5 for identities concerning covariant derivatives.

In this notation, the first order equations become (for $\lambda = 1$, $N > 0$)

$$g = i{*}g \quad , \qquad (1.7a,b),(6.4)$$

$$f = w \quad . \qquad (1.7c)\ ,(6.5)$$

Note some simple identities:

$$|\Phi|^2 = 1 - 2w \qquad (6.6)$$

$$\begin{aligned} dw &= -\frac{1}{2}(\bar{\Phi}d\Phi + \Phi d\bar{\Phi}) \\ &= -\frac{1}{2}(\bar{\Phi}D_A\Phi + \Phi D_A\bar{\Phi}) = -\frac{1}{2}(\Phi\bar{g} + \bar{\Phi}g) \\ &= -\operatorname{Re}(\Phi\bar{g}) \quad . \end{aligned} \qquad (6.7)$$

In particular,

$$-\bar{\Phi}g = (dw + i{*}df) \quad . \qquad (6.8)$$

Define $|g|^2 = \Sigma|g_k|^2 = {*}(g \wedge {*}\bar{g}) = (g,g)$.

PROPOSITION 6.1. *Let* (A,Φ) *be a smooth solution to the equations* (6.2-3). *Then*

$$\Delta w = -|g|^2 + \lambda(1 - 2w)w \qquad (6.9)$$

$$\Delta f = -i{*}(g \wedge \bar{g}) + (1 - 2w)f \qquad (6.10)$$

$$\Delta_A g = \frac{1}{2}(1+\lambda)(1-2w)g - \frac{1}{2}(1-\lambda)\Phi\Phi\bar{g} - if{*}g - \lambda wg \qquad (6.11)$$

$$\Delta_A^2 g = \frac{1}{2}(1+\lambda)(1-2w)g - \frac{1}{2}(1-\lambda)\Phi\Phi\bar{g} - 2if{*}g - \lambda wg \quad . \qquad (6.12)$$

Remark. The operator ∇_A^2 in (6.12) is defined by

$$\nabla_A^2 = \sum_k (\nabla_A)_k^2 \quad . \qquad (6.13)$$

Both ∇_A^2 and the covariant Laplacian

$$\Delta_A = {*}D_A{*}D_A + D_A{*}D_A{*} \qquad (6.14)$$

are gauge covariant and leave invariant the space of p-forms. They are related in the formulas (6.11-12) by

$$\Delta_A = \nabla_A^2 + if* \qquad d=2,\ p=1 \ . \tag{6.15}$$

Note that with our conventions, $\Delta_{A=0} = \Delta = \nabla^2$ is a negative operator. This identity is derived in Chapter VI, cf. Formula (5.11).

COROLLARY 6.2. *Let* (A,Φ) *be a smooth solution to* (6.2-3). *Then*

$$|g|\Delta|g| \geq -(2|f| + \lambda w)|g|^2 + [\min(1,\lambda)]\cdot(1-2w)|g|^2 \ .$$

For $\lambda = 1$,

$$|g|\Delta|g| \geq -(2|f| + w)|g|^2 + (1-2w)|g|^2 \ . \tag{6.16}$$

We now proceed to the proofs, which merely involve calculations.

<u>Proofs</u>. Using (6.7), and writing D in place of D_A,

$$\begin{aligned}\Delta w = *d*dw &= -\frac{1}{2} * \left\{ d\bar{\Phi} \wedge *g + d\Phi \wedge *\bar{g} + \bar{\Phi} d*g + \Phi d*\bar{g} \right\} \\ &= -\frac{1}{2} * \left\{ \overline{D\Phi} \wedge *g + D\Phi \wedge *\bar{g} + \bar{\Phi} D*g + \Phi \overline{D*g} \right\} \\ &= -|g|^2 - \mathrm{Re}(\bar{\Phi} * D * D\Phi) \ .\end{aligned} \tag{6.17}$$

Insert (6.3) into (6.17) to obtain (6.9).

The proof of (6.10) proceeds by applying $*d$ to the equation (6.2), namely

$$\begin{aligned}\Delta f = *d*df &= -\frac{i}{2} *d\{\Phi\bar{g} - \bar{\Phi} g\} \\ &= -\frac{i}{2} * \left\{ d\Phi \wedge g - d\bar{\Phi} \wedge g + \Phi d\bar{g} - \bar{\Phi} dg \right\} \\ &= -\frac{i}{2} * \Big\{ D\Phi \wedge g - \overline{D\Phi} \wedge g + i\Phi A \wedge d\bar{\Phi} + i\bar{\Phi} A \wedge d\Phi + i\Phi d(A\Phi) \\ &\quad + i\,\bar{\Phi} d(A\Phi) \Big\} = -i*(g \wedge \bar{g}) + f|\Phi|^2 \ .\end{aligned}$$

To establish (6.11), begin from (6.3) and differentiate to obtain

$$D{*}D{*}g \;=\; D{*}D{*}D\Phi \;=\; -\lambda D(\Phi w) \;=\; -\lambda(dw)\Phi - \lambda wg \,. \tag{6.18}$$

Likewise use

$$\begin{aligned} DD\Phi &= d(d-iA)\Phi - iA\wedge(d-iA)\Phi \\ &= -iF\Phi \end{aligned} \quad ,$$

to calculate

$$\begin{aligned} {*}D{*}Dg &= {*}D{*}D^2\Phi \;=\; -i{*}D{*}(F\Phi) \\ &= -i{*}\{\Phi df + fD\Phi\} \\ &= -i{*}\{\Phi df + fg\} \end{aligned} \quad . \tag{6.19}$$

Adding (6.18-19), we use (6.8) to give

$$\begin{aligned} \Delta_A g &= -\Phi\{\lambda dw + i{*}df\} - \lambda wg - if{*}g \\ &= \frac{1}{2}(1+\lambda)|\Phi|^2 g - \frac{1}{2}(1-\lambda)\Phi^2\bar{g} - if{*}g - \lambda wg \end{aligned} \quad .$$

This is (6.11).

To derive the corollary, we avoid the lack of regularity of $|g|$ at its zeros by defining

$$|g|_\rho^2 \;=\; |g|^2 + \rho^2 \quad .$$

Notice the elementary identity

$$\Delta|g|_\rho \;=\; |g|_\rho^{-1}\left\{\mathrm{Re}(\bar{g},\nabla_A^2 g) + |\nabla_A g|^2 - |g|_\rho^{-2}|\mathrm{Re}(\bar{g},\nabla_A g)|^2\right\} . \tag{6.20}$$

Multiply (6.20) by $|g|_\rho$ and let $\rho\to 0$. Since $g/|g|_\rho$ has a length bounded by one and converges for all g as $\rho\to 0$, each term on the right converges pointwise. Note, however, that by the Schwarz inequality, the sum of the last two terms in (6.20) is positive, i.e.

$$|g|_\rho^{-2}|(\bar{g},\nabla_A g)|^2 \le |\nabla_A g|^2 \quad .$$

Thus by (6.20), we obtain

$$|g|\Delta|g| \geq \mathrm{Re}(\bar{g},\nabla_A^2 g) \quad , \tag{6.21}$$

which is sometimes called Kato's inequality in unintegrated form, cf. Chapter VI.6. Substituting (6.12) for $\nabla_A^2 g$, we obtain (6.16) by using

$$|2if(g,*g)| \leq 2|f|\ |g|^2 \quad ,$$

$$|\Phi\Phi\bar{g}| \leq |\Phi|^2|g| \quad .$$

III.7. A METHOD TO ESTABLISH EXPONENTIAL DECAY

Before discussing the abelian Higgs model, we present a method for proving exponential decay. We apply the method in the next two sections. Basically, the method can be understood as follows: If $\Delta u = u$, then the fundamental solutions behave asymptotically as $\exp(\pm|x|)$ as $|x| \to \infty$. If we require some condition to ensure $u \to 0$ as $|x| \to \infty$, then the exponential increasing mode is ruled out and $u \leq \exp(-|x|)$. Actually the inequality $\Delta u \geq u$, for positive u is sufficient to conclude exponential decrease, once we know $u \to 0$. We use the maximum principle, which we state and prove in Chapter VI.3, to establish exponential decay.

In this section we deal with n-component vectors $u = (u_1,\dots,u_n) \in \mathbb{C}^n$, and ask under what conditions there exist m, M_1 with

$$|u| = \sum_{j=1}^{n} |u_j|^2 \leq M_1 e^{-m|x|} \quad . \tag{7.1}$$

We begin with an elementary inequality.

LEMMA 7.1. *Let* $u = (u_1 \dots u_n)$ *be a smooth vector-valued function on* $\mathbb{R}^d$. *Then* $|u|\Delta|u|$ *is continuous almost everywhere and for* u *real*

$$|u|\Delta|u| \geq (u,\nabla^2 u), \qquad \text{a.e.} \tag{7.2}$$

Proof. See the proof of (6.21), and Chapter VI.5.

PROPOSITION 7.2. *Let* $u = (u_1,\ldots,u_n)$ *be a smooth vector valued function on* $\mathbb{R}^d$ *such that*

$$\lim_{|x|\to\infty} |u| = 0 \tag{7.3}$$

$$|u|\Big|_{|x|=R} \le M \tag{7.4}$$

$$|u|\Delta|u| \ge m^2|u|^2 \tag{7.5}$$

where m, M, R *are constants. Then*

$$|u(x)| \le M_1 e^{-m|x|}, \qquad |x| > R, \tag{7.6}$$

where

$$M_1 = Me^{mR}. \tag{7.7}$$

Proof. Let $v(x) = M_1 \exp(-m|x|)$. Then

$$\Delta v = (-m(d-1)|x|^{-1} + m^2)v \le m^2 v \tag{7.8}$$

and $v(x) > 0$. It follows that

$$|u|\Delta(v-|u|) \le m^2|u|(v-|u|) \tag{7.9}$$

furthermore we have $v - |u|\Big|_{|x|=R} \ge 0$. Let $V = \{|x| > R$ and $|u(x)| > |v(x)|\}$. Since v is nonvanishing, $0 < |u| \in C^\infty(\bar{V})$. By taking a slightly larger open set V' if necessary we are assured that

(1) $\partial\bar{V}'$ is smooth and $\bar{V}'\cap\{|x| = R\} = \emptyset$,

(2) $0 < |u| \in C^\infty(\bar{V}')$,

(3) $v - |u|\Big|_{\partial\bar{V}'} \ge 0$.

Therefore we can divide both sides of (7.9) by $|u|$ in $\bar{V}'$, and $|u|$ satisfies

$$\Delta(v - |u|) \le m^2(v - |u|), \qquad x \in \bar{V}' \quad . \tag{7.10}$$

Applying the maximum principle in $\bar{V}'$ we conclude that $v - |u|\,\big|_{\bar{V}} \ge 0$. Since by definition, $v - |u| \ge 0$ in $\{|x| > R\} \setminus V$, the proposition follows.

COROLLARY 7.3. *Let* $u = (u_1, \ldots, u_n) \in C^\infty$ *satisfy* (7.3) *and* (7.4). *Let* $0 \le v$ satisfy

$$\lim_{R\to\infty} \sup_{|x|=R} v(x) = 0 \quad . \tag{7.11}$$

Suppose further that

$$|u|\Delta|u| \ge m^2|u|^2(1 - v) \quad . \tag{7.12}$$

Then for any $\varepsilon > 0$, *there exists* $M(\varepsilon) < \infty$ *such that*

$$|u(x)| \le M(\varepsilon)e^{-(1-\varepsilon)m|x|} \quad . \tag{7.13}$$

<u>Proof</u>. By the uniform convergence of v to zero, (7.11), we choose $R(\varepsilon)$ sufficiently large so that for $|x| \ge R(\varepsilon)$, we have $|v| \le \varepsilon/2$. By (7.12),

$$|u|\Delta|u| \ge m^2\left(1 - \frac{\varepsilon}{2}\right)|u|^2 \ge \left[\left(1 - \frac{\varepsilon}{2}\right)m\right]^2|u|^2 \quad .$$

Hence we infer (7.13) from the proposition.

In the next proposition we consider the effect of an exponentially decaying, inhomogeneous term.

PROPOSITION 7.4. *Let* $u \in C^\infty(\mathbb{R}^d)$, $v \in C^0(\mathbb{R}^d)$. *Suppose that for* $|x| \ge R$,

$$\Delta u - 2b^k \cdot \nabla_k u - m^2 u(1 - v) = q(x) \tag{7.14}$$

and

$$\lim_{|x|\to\infty} |u|, |v| \to 0 \quad \textit{uniformly in } |x| \quad . \tag{7.15}$$

Here b^k *is bounded and*

$$|q(x)| \le M_1 e^{-\sigma|x|}$$

for

$$0 < M_1 , \quad \sigma < \infty \quad . \tag{7.16}$$

Set $b_0 = \sup_{|x|>R} |b|$. *Then given* $0 < \varepsilon < 1$, *there exists* $M(\varepsilon) < \infty$ *such that*

$$|u| \le M(\varepsilon) e^{-(1-\varepsilon)\bar{m}|x|} \quad ,$$

with

$$\bar{m} = \min((m^2 + b_0^2)^{1/2} - b_0, \sigma) \quad . \tag{7.17}$$

Proof. Choose $R(\varepsilon) > R$ so that $\sup_{|x|>R(\varepsilon)} |v(x)| < \varepsilon/2$. Define a comparison function

$$s(x) = \frac{4\alpha}{\varepsilon\bar{m}^2} M_1 e^{-(1-\varepsilon)\bar{m}|x|} \quad , \tag{7.18}$$

where $\alpha > 1$ is to be determined below. Then (with possibly a new $R(\varepsilon)$)

$$\begin{aligned} \Delta s(x) &\le \bar{m}^2(1-\varepsilon/2)s(x) - \alpha|q| \quad , \quad |x| > R(\varepsilon) \\ -2b^k\nabla_k s(x) &= 2(1-\varepsilon)\bar{m}\, b^k \frac{x^k}{|x|} s(x) \end{aligned} \tag{7.19}$$

and

$$\Delta s - 2b^k\nabla_k s - m^2(1-\varepsilon/2)s \le -\alpha|q| \quad , \quad |x| > R(\varepsilon) \quad . \tag{7.20}$$

Because $\alpha > 1$, we obtain for $|x| > R(\varepsilon)$,

$$\Delta(s \pm u) - 2b^k\nabla_k(s \pm u) - m^2(1 - m^2(1-v)(s \pm u) \le 0 \quad .$$

For α sufficiently large,

$$s \pm u \Big|_{|x|=R(\varepsilon)} > 0 \quad . \tag{7.21}$$

Now use the maximum principle, Proposition VI.3.3

To end this section, we determine criteria for uniform decay. We establish the uniform decay of all functions u in an appropriate Sobolev space.

PROPOSITION 7.5. *Suppose* $u \in L^1_p$, *for some* $p > d$, *as defined in* VI.1. *Then* u *satisfies*

$$\lim_{R\to\infty} \sup_{|x|=R} |u(x)| = 0 \quad . \tag{7.22}$$

Proof. Suppose (7.22) did not apply. Then, there would exist an $\varepsilon > 0$ and a sequence of points $\{x_j\}$, $|x_j| \to \infty$, with $|u(x_j)| > \varepsilon$. Without loss of generality, assume $|x_i - x_j| > 2$. By the Corollary VI.2.7, there exists $\alpha > 0$ and $K < \infty$ such that u is Hölder continuous with exponent α, in the sense that for all x, y,

$$|u(x) - u(y)| \leq K|x-y|^{\alpha} \quad . \tag{7.23}$$

Let $\rho = \min(1,\varepsilon/2K)^{1/\alpha}$. By (7.23), we infer that for all $|y - x_j| < \rho$,

$$|u(y)| = |u(x_j) + u(y) - u(x_j)| \geq \varepsilon/2 \quad .$$

This yields a divergent estimate on the L_ρ norm of u, namely

$$\int_{R^d} |u|^p \, dy \geq \sum_{j=1}^{\infty} \int_{|y-x_j|<\rho} |u(y)|^p \, dy \geq \sum_j \left(\frac{\varepsilon}{2}\right)^p \Omega(d)\rho^d$$

$$= \infty ,$$

where $\Omega(d)$ is the volume of the unit d-ball. Since $L^1_p \subset L_p$, the proposition is proved.

III.8. EXPONENTIAL DECAY (MAGNETIC SCREENING)

In this section we establish exponential decay for $1-|\Phi|^2$ and for $D_A\Phi$ whenever (A,Φ) is a smooth, finite action solution to the second order equations for the abelian Higgs model with $\lambda > 0$. This exponential decay is called the "Higgs effect" since one might *a priori* expect the $U(1)$ symmetry of Φ to produce a polynomial decay mode, arising from the one parameter family of minima of the potential $(|\Phi|^2-1)^2$. However, the $U(1)$ symmetry of the gauge field forbids the existence of the long range force, leaving an exponential decay. The fundamental reason that $w = \frac{1}{2}(1-|\Phi|^2)$ decays can be understood by inspecting the following equation, derived as (6.9),

$$\Delta w - \lambda w = -|g|^2 - 2\lambda w^2 \quad . \tag{8.1}$$

The basic idea is that if $|g| \le \exp(-\alpha|x|)$, then the results of §7 show that $w = \exp(-\lambda^{\frac{1}{2}}|x|)$, as long as $2\alpha \le \lambda^{\frac{1}{2}}$. We state precise decay estimates:

THEOREM 8.1. *Let* (A,Φ) *be a finite action, smooth solution to* (6.2-3) *or* (1.4). *Then either* $|\Phi| \equiv 1$, *or else*

$$|\Phi(x)| < 1 \quad . \tag{8.2a}$$

If $\lambda \le 1$,

$$|f| = |{*F}| \le \frac{1}{2}(1-|\Phi|^2) \quad , \tag{8.2b}$$

$$|D_A\Phi| \le \frac{3}{2}(1-|\Phi|^2) \quad . \tag{8.2c}$$

Hence if $|\Phi| \equiv 1$, *we infer* $F \equiv 0$, $D_A\Phi \equiv 0$.

For every $\lambda > 0$, *given* $\varepsilon > 0$, *there exists* $M = M(\varepsilon,\lambda) < \infty$ *such that*

$$|\mathrm{Re}(\bar{\Phi}D_A\Phi)|,\ (1-|\Phi|^2) \le Me^{-(1-\varepsilon)m_L|x|} \quad , \tag{8.3a}$$

and also

$$|\mathrm{Im}(\bar{\Phi}D_A\Phi)|,\ |{*F}| \le Me^{-(1-\varepsilon)|x|} \quad . \tag{8.3b}$$

Here $m_L = \min(\lambda^{1/2}, 2)$ *is the Higgs (longitudinal) mass.*

COROLLARY 8.2. *The vortex number of any smooth, finite action solution to* (1.4) *for* $\lambda > 0$ *as defined by* (1.3) *is integer.*

The proof of Theorem naturally divides into three parts. The first two establish (8.2), the last (8.3).

LEMMA 8.3. *Let* (A, Φ) *be a smooth solution to the second order equations* (6.2-3), *and satisfy* $w = \frac{1}{2}(1 - |\Phi|^2) \in L_2(\mathbb{R}^2)$. *Then either* $|\Phi| \equiv 1$, *or else* $|\Phi(x)| < 1$.

Proof. By (6.9), and the fact that $0 \leq 1 - 2w = |\Phi|^2$ it follows that $\Delta w \leq \lambda(1-2w)w$. Apply the maximum principle of Chapter VI.3 with $V = \mathbb{R}^2$.

LEMMA 8.4. *Let* (A, Φ) *be a smooth solution to the second order equations* (6.2-3) *with* $\lambda \leq 1$. *Assume finite action. Then* (8.2b) *and* (8.2c) *hold.*

Proof. We first establish (8.2b). Adding equations (6.9) and (6.10) we obtain (for $\lambda \leq 1$)

$$\Delta(w + f) \leq -|g|^2 + |\bar{g} \wedge g| + (1 - 2w)(\lambda w + f) \leq (1 - 2w)(w + f) \tag{8.4}$$

where we have used Schwarz' inequality to conclude that

$$|\bar{g} \wedge g| \leq |g|^2 \quad . \tag{8.5}$$

Applying the maximum principle (see Chapter VI.3) to (8.4) with $V = \mathbb{R}^2$, we conclude $w \geq -f$. Next subtract equation (6.10) from (6.9) to obtain

$$\Delta(w - f) \leq -|g|^2 + |\bar{g} \wedge g| + (1 - 2w)(\lambda w - f) \leq (1 - 2w)(w - f) \ . \tag{8.6}$$

Here again, (8.5) is used. The maximum principle and (8.6) imply $w \geq +f$, proving statement (8.2b).

As for (8.2c), we write this inequality as $|g| \le 3w$. Multiply the identity (6.9) by $3|g|$ and subtract (6.16). Thus

$$|g|\Delta(3w - |g|) \le |g|(3w - |g|)(1 - 2w + |g|) - 2|g|^3$$
$$\le |g|(3w - |g|)(1 - 2w + |g|) \quad . \tag{8.7}$$

Here we have used the fact that $|f| \le w$ and $\lambda \le 1$. By Corollary II.2.3, assume $0 < w(x)$. Thus $3w = |g|$ only at points where $|g| \ne 0$. Let $V_1 = \{x$ where $3w - |g| < 0\}$. Since g is smooth, we can enlarge V_1 to V such that V is open, $\partial\bar{V}$ is smooth, $|g| \ne 0$ on $\bar{V}$. Hence $3w - |g| \ge 0$ on $\partial\bar{V}$. Furthermore since $|g| \ne 0$, we divide (8.7) to obtain

$$\Delta(3w - |g|) \le (3w - |g|)(1 - 2w + |g|) \quad .$$

Since $1 - 2w + |g| > 0$, the maximum principle applies, it yields $3w - |g| \ge 0$, as desired.

The proof of exponential decay (8.3) is technically simpler for $\lambda = 1$. We consider this case separately and treat $\lambda \ne 1$ in the next section. In outline, the two proofs are similar.

THEOREM 8.5. *Let* (A,Φ) *be a smooth solution to the first order equations* (1.7). *Then for any* $\varepsilon > 0$, *there exists* $M(\varepsilon) < \infty$ *such that*

$$0 < w = \frac{1}{2}(1 - |\Phi|^2) < M(\varepsilon)e^{-|x|(1-\varepsilon)} \tag{8.8}$$

Proof. Since $0 < w \le 1/2$, and $w \in H_1 \subset L_2$, it follows that $w \in L_p$ for all $p \ge 2$. We show that $dw \in L_p$, $p > 2$. In fact, by (6.7),

$$dw = -\mathrm{Re}(\bar{\Phi}g) \quad . \tag{8.9}$$

By (8.2a), $|\Phi| < 1$, so it is sufficient to prove $g \in L_p$. By Theorem 8.1, $|g| \le 3w$. But we just remarked that $w \in L_p$ for all $p \ge 2$. In particular, $w \in L_p^1$ for all $p \ge 2$.

We now appeal to Proposition 7.5, namely given $\varepsilon > 0$, there exists $R(\varepsilon)$ such that

$$0 < w < \varepsilon, \qquad \text{if} \quad R(\varepsilon) < |x| \quad . \tag{8.10}$$

Thus $|g|^2 \leq 9w^2$, and by the identity (6.6);

$$\Delta w \geq (1 - 11w)w \geq (1 - \varepsilon)w \quad . \tag{8.11}$$

Now apply Proposition 7.2.

III.9 THE HIGGS MASS AND THE PHOTON MASS

For $\lambda \neq 1$, the Higgs and photon masses differ. The Higgs mass is λ-dependent and arises from the coupling $\lambda(|\Phi|^2 - 1)^2/8$. The photon mass arises from $\frac{1}{2}|A|^2|\Phi|^2$ and equals one. We also prove here that $m_{\text{Higgs}} \leq 2\, m_{\text{photon}}$.

THEOREM 9.1. *Let* (A,Φ) *be a smooth, finite action solution to the second order equations* (1.4). *Then the decay* (8.3) *of Theorem* 8.1 *holds.*

LEMMA 9.2. *With* f,g,w *as before,*

$$f,g,w \in L_p^1 \, , \qquad \textit{all} \quad 2 \leq p < \infty \quad . \tag{9.1a}$$

$$\lim_{R\to\infty} \sup_{|x|=R} \{|f|, |g|, w\} = 0 \quad . \tag{9.1b}$$

Proof. By Proposition 7.5, the estimate (9.1b) follows from (9.1a). By assumption, $\{f,g,w\}$ are L_2. By Theorem 8.1, $|\Phi| < 1$. Using (6.1) and (6.7),

$$|\nabla f| = \frac{1}{2} |\Phi\bar{g} - \bar{\Phi}g| \leq |g| \quad , \tag{9.2}$$

$$|\nabla w| = \frac{1}{2} |\Phi\bar{g} + \bar{\Phi}g| \leq |g| \quad . \tag{9.3}$$

Hence $f, w \in L_2^1$. By a Sobolev inequality (Proposition VI.2.4), $f, w \in L_p$ for $2 \le p < \infty$. Now consider g. Write equation (6.12) in the form

$$\nabla_A^2 g = \Lambda(g) + v \quad , \tag{9.4}$$

where $\Lambda(g) = -2if*g$ and

$$v = \frac{1}{2}(1+\lambda)(1-2w)g - \frac{1}{2}(1-\lambda)\Phi\Phi\bar{g} - \lambda wg \in L_2 \; .$$

We refer forward to Theorem V.7.1, which ensures

$$g \in L_p^1 \; , \qquad \text{all} \quad 2 \le p < \infty \quad . \tag{9.5}$$

Returning to (9.2-3), we observe that f and w are also L_p^1 for all $p < \infty$ as desired.

To proceed with the proof of the theorem, decompose the equation (9.4) into separate equations for $\mathrm{Re}(\bar{\Phi}g)$ and $\mathrm{Im}(\bar{\Phi}g)$. It is easiest to work in a gauge in which $\Phi > 0$. The resulting differential inequalities that we establish, however, are gauge-independent. Let $V_R = \mathbb{R}^2 \setminus \{x : |x| \le R\}$. The lemma ensures that for R sufficiently large, $|\Phi| \ge \frac{1}{2}$. Thus the function $U_R = \bar{\Phi}|\Phi|^{-1}$ is C^∞ on V_R. This defines a gauge transformation U_R, for which $U_R\Phi > 0$.

On V_R, let $g = g_1 + ig_2$ denote the decomposition of g into real and imaginary parts. Then (9.4), (6.9-10) can be written

$$\nabla_A^2 g = \lambda(1-3w)g_1 + 2f*g_2 + i((1-2w-\lambda w)g_2 - 2f*g_1) \tag{9.6a}$$

$$\Delta f = 2*(g_2 \wedge g_1) + (1-2w)f \quad , \tag{9.6b}$$

$$\Delta w = -|g_1|^2 - |g_2|^2 + \lambda(1-2w)w \quad . \tag{9.6c}$$

Because ∇_A^2 is not real, equation (9.6a) couples g_1 and g_2. In our gauge, $g_1 = d\Phi$, $g_2 = -A\Phi$. By (1.4a), $d*(\Phi g_2) = 0$. Thus

$$\nabla_A^2 g - \Delta g = 2i\Phi^{-1} g_{2k}\nabla_{A_k} g + ig{*}d{*}(\Phi^{-1}g_2) + \Phi^{-2}|g_2|^2 g \quad . \tag{9.7}$$

Define $\psi = \begin{pmatrix} g_2 \\ f \end{pmatrix}$. From (9.6-7) we obtain

$$-\Delta\psi + \psi = \Lambda_1(\psi) \ , \tag{9.8}$$

where the nonlinear terms in (9.8) are given by the linear transformation Λ_1;

$$\Lambda_1(\psi) = \begin{pmatrix} (2+\lambda)wg_2 + 2f{*}g_1 + 2\Phi^{-1}\{g_{2k}\nabla_k g_1 + g_{2k}A_k g_2\} - 2\Phi^{-2}(g_2,g_1)g_1 + \Phi^{-2}|g_2|^2 g_2 \\ 2wf + 2{*}(g_1 \wedge g_2) \end{pmatrix} \tag{9.9}$$

Note Λ_1 is well-defined, since no product of f with g_2 occurs in (9.9). The vector ψ, as a section of $T^* \oplus \mathbb{R}$, inherits a natural inner product,

$$\left(\begin{pmatrix} a \\ b \end{pmatrix}, \begin{pmatrix} c \\ d \end{pmatrix}\right) = *(a \wedge *c) + bd \quad .$$

Let $|\Lambda|$ denote the (pointwise) operator norm of Λ with respect to this inner product space. By (9.8),

$$(\psi,\Delta\psi) = |\Phi|^2 - (\psi,\Delta(\psi)) \geq |\Phi|^2(1 - |\Lambda|) \quad .$$

By Lemma 7.1,

$$(|\psi|,\Delta|\psi|) \geq |\psi|^2(1 - |\Lambda|) \quad . \tag{9.10}$$

in V_R. As $|\Lambda|$ and $|\Psi|$ are gauge-independent, (9.10) holds in all gauges.

An estimate

$$|\psi| \leq M(\varepsilon)e^{-(1-\varepsilon)|x|} \tag{9.11}$$

would ensure the desired inequality (8.3b). We show below that $|\Lambda| \to 0$ uniformly as $|x| \to \infty$. Since also w, $|f|$, $|g| \to 0$ (Lemma 9.2), it follows by Corollary 7.3 that for any $\varepsilon > 0$, there exists $M(\varepsilon) < \infty$ such that (9.11) holds.

LEMMA 9.3.

$$\lim_{R\to\infty} \sup_{|x|=R} \{|\nabla_A g|, |\Lambda|\} = 0 \quad . \tag{9.12}$$

Proof. First we reduce the $|\Lambda|$-bound to the bound on $\nabla_A g$. In the gauge we are using,

$$A = -\Phi^{-1} g_2, \qquad \text{so} \qquad |A| \le 2|g_2| \quad . \tag{9.13}$$

By Lemma 9.2, $|A| \to 0$ uniformly. Furthermore,

$$\nabla_k g = \nabla_{A_k} g + iA_k g \quad ,$$

Again by Lemma 9.2, if $|\nabla_A g| \to 0$, then $|\nabla_k g| \to 0$. It therefore follows by one more application of (9.1) that $|\Lambda| \to 0$ uniformly. Thus we need only bound $\nabla_A g$.

It is sufficient to prove that $\nabla_A g \in L_p^1$ for some $p > 2$ (Proposition 7.5). To establish this, we derive an equation for $\nabla_A g$ by applying ∇_{A_j} to (9.4). The commutator $[\nabla_A^2, \nabla_{A_j}]$ is evaluated with the help of Proposition VI.5.2 to give

$$\nabla_A^2 (\nabla_{A_j} g) = v_j \quad , \tag{9.14}$$

where

$$\begin{aligned} v_j = {} & -2if*(\nabla_{A_j} g) + 2i\varepsilon_{jk} f \nabla_{A_k} g + i\varepsilon_{jk}(\nabla_k f) g - 2i(\nabla_j f)*g \\ & + \frac{1}{2}(1+\lambda)(1-2w)\nabla_{A_j} g - \frac{1}{2}(1-\lambda)\Phi\Phi\overline{\nabla_{A_j} g} - \lambda w \nabla_{A_j} g \\ & - ((1+2\lambda)\nabla_j w) g - (1-\lambda)\Phi\bar{g}\nabla_{A_j}\Phi \quad . \end{aligned} \tag{9.15}$$

We claim $v_j \in L_2$, which can be justified by showing each of the nine terms in (9.15) is L_2. By Lemma 9.2, f and $\nabla_{A_j} g$ are both L_4, so their product is L_2. The bound on the second, third and fourth terms is similar. The remaining terms are products of L_∞ functions with either $\nabla_{A_j} g \in L_2$ or $g \in L_2$, and hence are L_2. Since $v_j \in L_2$, we appeal to Theorem V.7.1, with $\Lambda = 0$, to infer that $\nabla_A g \in L_p^1$ for $2 \le p < \infty$.

At this point we have established (9.11) and have obtained the decay of $|g_2|$, $|f|$. In order to complete the proof of Theorem 9.1, we need to establish decay of g_1, w.

Using the real part of (9.6a),

$$\Delta g_1 - \lambda(1 - 3w) g_1 - \Phi^{-2} |g_2|^2 g_1 + 2\Phi^{-2} (g_{1k} g_{2k}) g_2$$

$$= \quad 2f * g_2 + 2\Phi^{-1} g_{2k} \nabla_k g_2 \quad . \tag{9.16}$$

We have established uniform decay of w and $|g_2|$ as $|x| \to \infty$. Thus (9.16) is an equation of the form treated in Proposition 7.4. The right hand side of (9.16) is bounded by $M \exp(-(1-\varepsilon)|x|)$. Hence by Proposition 7.4,

$$|g_1| \le M(\varepsilon) \exp(-(1-\varepsilon)\bar{m}|x|) \quad , \tag{9.17}$$

where $\bar{m} = \min(\lambda^{1/2}, 1)$. Note that $\bar{m}$ is not the mass $m_L = \min(\lambda^{1/2}, 2)$ of Theorem 8.1.

Next we prove that w decays with mass m_L. Recall that w satisfies (9.6c). Thus the decay for $|g|$ already proven in (9.11), (9.17) yields, by Proposition 7.4,

$$|w| \le M(\varepsilon) \exp(-(1-\varepsilon) m_L |x|) \quad . \tag{9.18}$$

Again using (9.6c),

$$(-\Delta + \lambda) w \quad = \quad |g|^2 + 2\lambda w^2 \quad .$$

Let $G_\lambda(x-y)$ denote the Green's function of $(-\Delta+\lambda)$ (see (13.11) of Chapter IV). Then

$$\nabla w(x) = \int G_\lambda(x-y)\nabla(|g|^2(y) + 2\lambda w^2(y))\, dy \quad . \tag{9.19}$$

Note $G_\lambda(x-y)$ is locally integrable, and for $|x|$ bounded away from zero,

$$G_\lambda(x) \le \text{const } e^{-\lambda^{\frac{1}{2}} x} \quad .$$

The convolution of G_λ with a smooth function bounded by $\exp(-m|x|)$ yields a function which decreases exponentially with rate $\min(\lambda^{\frac{1}{2}}, m)$. Hence by (9.11), (9.17-19) it follows that for $\varepsilon > 0$, there exists $M(\varepsilon)$ such that

$$|\nabla w| \le M(\varepsilon) \exp(-(1-\varepsilon)m_L|x|) \quad . \tag{9.20}$$

Recall that for $|x|$ sufficiently large,

$$|g_1| = |\nabla\Phi| \le 2|\nabla w| \quad .$$

Thus

$$|g_1| \le M(\varepsilon) \exp(-(1-\varepsilon)m_L|x|)$$

as desired. This completes the proof of Theorem 9.1.

III.10 EQUIVALENCE OF FIRST AND SECOND ORDER EQUATIONS

THEOREM 10.1. *Every critical point of the functional with* $\lambda = 1$, *i.e. of*

$$\mathscr{A}(A,\Phi) = \frac{1}{2}\int *(f^2 + |g|^2 + w^2) \quad , \tag{10.1}$$

is a minimum of $\mathscr{A}$ *with fixed* $N \equiv (2\pi)^{-1}\int F \in \mathbb{Z}$.

Remark. The critical points of $\mathscr{A}$ are finite action solutions to the second order equations (6.2-3), i.e. (1.4), with $\lambda = 1$. The minima of $\mathscr{A}$, for given integer* N, are finite action solutions of the first order equations (6.4-5) or (1.7.8). Thus the theorem ensures that every finite action solution of the second order equations with $\lambda = 1$ is a solution of the first order equations. Clearly the reverse is true, so the first and second order equations are equivalent, at least for finite action. (In fact we do not expect the equivalence for solutions with singularities.)

Proof. Let us define

$$u_+ = w + f , \qquad u_- = w - f . \tag{10.2}$$

Recall the first order equations to prove are

$$*g = -ig, \quad u_- = 0 \qquad \text{if } N > 0 , \tag{10.3}$$

$$*g = ig , \quad u_+ = 0 \qquad \text{if } N < 0 . \tag{10.4}$$

Also recall that by Theorem 8.1, a smooth solution of the second order equations with $\lambda = 1$ yields

$$u_+ , \; u_- \geq 0 . \tag{10.5}$$

By Corollary II.2.2 with $\lambda = 1$, $d = 2$, we have

$$\int_{\mathbb{R}^2} f^2 = \int_{\mathbb{R}^2} w^2 , \tag{10.6}$$

or

$$\int_{\mathbb{R}^2} u_+ u_- = 0 . \tag{10.7}$$

---

*It is not known at present whether $\mathscr{A}(A,\Phi) < \infty$ ensures $N \equiv 1/2\pi \int F \in Z$. That statement would mean that the function space defined by the condition $\mathscr{A} < \infty$ is divided into a set of discrete components characterized by the vortex (first Chern) number. It is proved in §§8 and 9, however, that every critical point of $\mathscr{A}$ has integer vortex number. Thus minimizing $\mathscr{A}$ with N integer recovers all critical points of $\mathscr{A}$.

Thus for every x, either $u_+(x) = 0$ or $u_-(x) = 0$. In other words

$$\pm f(x) = w(x) \quad , \tag{10.8}$$

But by Theorem 8.1, either $w \equiv 0$ (in which case $f \equiv 0$, $|\Phi| \equiv 1$, $N = 0$, and in an appropriate gauge $A = 0$, $\Phi = 1$) or else $w > 0$. The latter assures that f has constant sign, since f is smooth and has no zero by virtue of (10.8). We consider the two cases (10.8), with the same sign for every x, and establish (10.3) or (10.4).

Recall the identities (6.9-10) with $\lambda = 1$,

$$\Delta w = -*(g \wedge *\bar{g}) + (1 - 2w)w \tag{10.9}$$

$$\Delta f = -i*(g \wedge \bar{g}) + (1 - 2w)f \quad . \tag{10.10}$$

If $f = w$, i.e. $N > 0$, add (10.9-10) to obtain

$$|g - i*g|^2 = *((g - i*g) \wedge (*\bar{g} + i\bar{g})) = 0 \ .$$

If $-f = w$, i.e. $N < 0$, add to obtain

$$|g + i*g|^2 = 0 \quad .$$

This proves (10.3 or 4). By Corollary 8.2, $N \in \mathbb{Z}$, and hence every critical point of $\mathscr{A}$, with $\lambda = 1$, is a solution to the first order equations given in Theorem 1.1. This completes the proof.

III.11 SELF-DUALITY FOR THE O(3) - SYMMETRIC YANG-MILLS THEORIES ON R^4

Because of the many formal similarities between the O(3)-symmetric, SU(2) Yang-Mills equations on $\mathbb{R}^4$ as defined in Chapter II and the abelian Higgs model with $\lambda = 1$ as defined in §1 here, it is not surprising that the results of this chapter have analogs. For the Yang-Mills theory, the minima of the action can be obtained in closed form--they are instantons or anti-instantons on a line. The question is, do there exist finite action, O(3) symmetric critical points to the SU(2) Yang-Mills functional on $\mathbb{R}^4$, which are neither instantons nor anti-instantons? The answer, under the mild regularity assumptions stated below, is no.

Before stating the result precisely as Theorem 11.1, we remind the reader of some results of Chapter II. An O(3) symmetric connection, $\hat{A}$, on $\mathbb{R}^4$ is equivalent to a U(1) connection $A(r,t)$ and a complex scalar field $\Phi(r,t)$ defined on the Poincaré half plane $\overline{\mathbb{R}}^2_+ = \{(r,t) \in \mathbb{R}^2 \mid r \geq 0\}$. The Riemannian metric on $\mathbb{R}^2_+$ is

$$ds^2 = \frac{1}{r^2}(dr^2 + dt^2) \quad . \tag{11.1}$$

Let $\hat{F}_{\hat{A}}$ denote the SU(2) curvature on $\mathbb{R}^4$. The action is

$$\mathscr{A} = \frac{1}{2}\langle \hat{F}_{\hat{A}}, \hat{F}_{\hat{A}}\rangle_{L_2(\mathbb{R}^4)} = \pi \int_{\mathbb{R}^2_+} \left\{ r^2 F_A \wedge *F_A + 2D_A\Phi \wedge *D_A\Phi + \frac{*1}{r^2}(|\Phi|^2 - 1)^2 \right\} \tag{11.2}$$

where

$$*dt = dr, \qquad *dr = -dt, \qquad *1 = dt \wedge dr \quad .$$

The O(3) – symmetric Yang-Mills equations are equivalent to the following equations on $\mathbb{R}^2_+$:

$$d*(r^2 F_A) - i*(\Phi\overline{D_A\Phi} - \bar{\Phi}D_A\Phi) = 0$$

$$-D_A * D_A\Phi + * \frac{1}{r^2}(|\Phi|^2 - 1)\Phi = 0 \tag{11.3}$$

The second Chern number of an O(3) – symmetric connection on $\mathbb{R}^4$ equals the first Chern number of A on $\mathbb{R}^2_+$:

$$N = \frac{1}{2\pi}\int_{\mathbb{R}^2_+} F_A = -\frac{1}{8\pi^2}\,\mathrm{trace}\int_{\mathbb{R}^4} \hat{F}_{\hat{A}} \wedge \hat{F}_{\hat{A}} \quad . \tag{11.4}$$

The equations for instantons and anti-instantons are similar to equations similar to (1.7) and (1.8), respectively:

$$\left.\begin{aligned} D_A\Phi - i*D_A\Phi &= 0 \\ r*F_A - \frac{1}{r}(1-\Phi\bar{\Phi}) &= 0 \end{aligned}\right\} \qquad \text{If} \quad N \geq 0 \tag{11.5a}$$

$$\left.\begin{aligned} D_A\Phi + i*D_A\Phi &= 0 \\ r*F_A + \frac{1}{r}(1-\Phi\bar{\Phi}) & \end{aligned}\right\} \qquad \text{If} \quad N \leq 0 \tag{11.5b}$$

We remark that for fixed $|N|$ there exists a $2|N|$ parameter family of solutions to (11.5a,b). This is the same dimension as that of the parameter space of solutions to (1.7) and (1.8) for fixed $|N|$. The interpretation of the parameters is different: $|N|$ parameters specify the positions on the $r = 0$ axis of the instantons, while the remaining $|N|$ parameters determine the instanton sizes.

Having completed this brief review, we state as Theorem 11.1 the equivalence of equations (11.5) and (11.3).

THEOREM 11.1. *Let* $\hat{A}$ *be a smooth* O(3) *symmetric connection on* $\mathbb{R}^4$ *as defined in Chapter* II. *Suppose that the action of* $\hat{A}$ *is finite and that* $\hat{A}$ *is a solution to the Yang-Mills equations on* $\mathbb{R}^4$. *That is, the fields* (A,Φ) *on* $\mathbb{R}^2_+$ *defined by* $\hat{A}$ *satisfy equation* (11.3). *Then if the integer* N *defined by* (11.4) *is positive,* (A,Φ) *satisfy* (11.5a). *If the integer* N *is negative,* (A,Φ) *satisfy* (11.5b). *If* $N = 0$, (A,Φ) *is a gauge transformation of the configuration* $(A = 0,\ \Phi = 1)$.

We remark that the requirement that $\hat{A}$ be smooth is not important. Recent work by K. Uhlenbeck implies that every locally L^1_2 solution to the Yang-Mills equations on $\mathbb{R}^4$ is smooth, see Chapter V.

The proof of Theorem 11.1 is conceptually the same as the proof of Theorem 10.1. The added technical obstacle is the fact that the metric is hyperbolic. Thus care is required near the boundary of $\mathbb{R}^2_+$. The proof specifically uses the fact that if $\hat{A}$ is smooth, then for some ε,

$$\langle r^{-1-\varepsilon} F_{\hat{A}}, F_{\hat{A}} \rangle_{L_2} < \infty \quad . \tag{11.6}$$

We refer the reader to the original paper for the technical details of the proof [6].

We end this section with one additional comment: The O(3) symmetric, SU(2) Yang-Mills equations on $\mathbb{R}^2 \times S^2$ are equivalent to the system of equations (1.4a,b) with $\lambda = 1$ (cf. Chapter II). The only solutions to the O(3) symmetric equations on $\mathbb{R}^2 \times S^2$ are instantons or anti-instantons.

IV. (Part I) MONOPOLE CONSTRUCTION

IV.1. THE NON-ABELIAN HIGGS MODEL

The non-abelian Higgs model on $\mathbb{R}^3$ has the action functional

$$\mathscr{A}(A,\Phi) = \frac{1}{2}\|F\|^2_{L_2} + \frac{1}{2}\|D_A\Phi\|^2_{L_2} + \frac{\lambda}{8}\|\,|\Phi|^2 - 1\|^2_{L_2} \quad . \tag{1.1}$$

In this case the connection A, as well as Φ, take values in a matrix representation of $\mathfrak{g}$, the Lie algebra of a non-abelian group G. The absolute value of Φ and the inner products (,) of forms involve a suitably normalized inner product on the Lie algebra, e.g.,

$$(\Phi,\Phi) = \mathrm{Tr}(\Phi(x)^{\dagger}\Phi(x)) \quad .$$

See Chapter I.1 for further discussion. Here $\Phi^{\dagger}$ is the Hermitian conjugate of the matrix $\Phi(x)$. The curvature F is related to A by

$$F_{ij} = \partial_i A_j - \partial_j A_i + [A_i, A_j] \quad , \tag{1.2}$$

and

$$D_A\Phi = d\Phi + [A,\Phi] \tag{1.3}$$

Critical points of the functional $\mathscr{A}(A,\Phi)$ satisfy the second order system of equations on $\mathbb{R}^3$,

$$*D_A{*F} = J = [D_A\Phi,\Phi] \quad , \tag{1.4a}$$

$$\nabla^2_A\Phi = *D_A{*D_A\Phi} = \frac{\lambda}{2}\Phi(|\Phi|^2 - 1) \quad . \tag{1.4b}$$

Finite action ensures that solutions to (1.4) satisfy the boundary conditions

$$|\Phi| \to 1 \quad , \tag{1.5a}$$

$$|D_A\Phi| \to 0 \quad , \tag{1.5b}$$

$$|F| \to 0 \quad , \tag{1.5c}$$

uniformly as $|x| \to \infty$, as proved in §§10-15.

In $d = 3$, as opposed to the $d = 2$ equations of Chapter III, the $\lambda = 0$ equations have considerable interest. In fact the case $\lambda = 0$ in $d = 3$ has many features in common with $\lambda = 1$ in $d = 2$.

For $\lambda = 0$, there is no reason to expect $|\Phi| \to 1$, but we now explain why the boundary conditions (1.5) are natural even when $\lambda = 0$. In fact we prove in §11 that every finite action critical point of (1.1) with $\lambda = 0$ has a uniform $|x| \to \infty$ limit: For some constant c,

$$\lim_{R\to\infty} \sup_{|x|=R} |\Phi(x)| = \lim_{R\to\infty} \inf_{|x|=R} |\Phi(x)| = c \quad . \tag{1.6}$$

Furthermore, for all x,

$$|\Phi(x)| \le c \quad . \tag{1.7}$$

In fact (A,Φ) is trivial if $c = 0$; see Corollary 10.4.

On the other hand if (1.6) holds for $c \neq 0$, define a scaled field (A',Φ') by

$$(A'(x),\Phi'(x)) = (c^{-1}A(c^{-1}x), c^{-1}\Phi(c^{-1}x)) \quad . \tag{1.8}$$

If (A,Φ) satisfies (1.4), (1.6), (1.5b,c), then (A',Φ') satisfies (1.4-5), and has action $\mathscr{A}(A',\Phi') = c^{-1}\mathscr{A}(A,\Phi)$. Thus we restrict attention for all $0 \le \lambda$ to the case of boundary conditions (1.5a-c). Every finite action critical point, $0 \le \lambda$, achieves the limits (1.5) sufficiently rapidly to define the invariant

$$N = \frac{1}{4\pi} \operatorname{Tr} \int_{\mathbb{R}^3} F \wedge D_A\Phi ,$$

$$= \frac{1}{4\pi} \operatorname{Tr} \int_{S^2_\infty} \Phi F . \tag{1.9}$$

The last equality follows from the Bianchi identity, yielding $\operatorname{Tr}(F \wedge D_A\Phi) = \operatorname{Tr} d(\Phi F)$, and from Stokes' theorem to transform to an integral over S^2_∞, the two-sphere at infinity. In the case of $G = SU(2)$, N is an integer and equals the winding number of the map Φ defined at infinity (see (II.5.7)):

$$N = -\frac{1}{4\pi} \operatorname{Tr} \int_{S^2_\infty} \Phi \, d\Phi \wedge d\Phi , \qquad G = SU(2) . \tag{1.10}$$

For the case of general G, see Chapter II.3. We call N the monopole number, in deferrence to the case (1.10).

The action $\mathscr{A}$ can be written as a sum of squares plus a multiple of N. In fact

$$\mathscr{A}(A,\Phi) = \frac{1}{2}\| {*F} \mp D_A\Phi\|^2 + \frac{\lambda}{8}\| |\Phi|^2 - 1\|^2 \pm 4\pi N . \tag{1.11}$$

As a consequence of these identities,

$$\mathscr{A}(A,\Phi) \geq 4\pi |N| . \tag{1.12}$$

Consider the equality:

$$\mathscr{A} = 4\pi |N| . \tag{1.13}$$

REMARK. *If* $\lambda > 0$, *then* (1.13) *has no solution except* $\mathscr{A} = N = 0$.

For if $\lambda > 0$, then (1.13) ensures $|\Phi(x)| \equiv 1$. By Corollary II.2.3, $|F(x)| + |D_A\Phi(x)| = 0$. Thus $\mathscr{A} = 0$.

It is not known whether for $\lambda > 0$ there are stable minima of $\mathscr{A}$ for fixed $|N| \geq 2$. On the other hand (1.13) can only be satisfied for $\lambda = 0$. In that case, the minima of $\mathscr{A}$ for given N satisfy the first order equations

$$*F = D_A\Phi \qquad N < 0 \quad , \tag{1.14a}$$

$$*F = -D_A\Phi \qquad N > 0 \quad . \tag{1.14b}$$

The equations (1.14) with the boundary conditions (1.5) are the monopole equations.

For simplicity we take $G = SU(2)$ throughout the remainder of this section. In the case $N = \pm 1$, an explicit solution to (1.14) is known, the Prasad-Sommerfield monopole.

For notation, let

$$r = |x| \quad , \qquad n_i = x_i/r \ , \qquad e_j = \frac{i}{2}\ \sigma_j \ .$$

Thus r denotes Euclidean distance, $\vec{n}$ is a unit radial vector, σ_j are Pauli matrices and the trace, Tr, is normalized so that $\{e_j\}$ is an orthonormal basis for the Lie algebra of $SU(2)$. The $N = \pm 1$ monopoles located at $r = 0$ are given by

$$\Phi(x) = \mp\left(\frac{1}{\tanh r} - \frac{1}{r}\right)\vec{n}\cdot\vec{e} \quad , \tag{1.15a}$$

$$A(x) = \left(\frac{1}{\sinh r} - \frac{1}{r}\right)(\vec{n}\times\vec{e})\cdot d\vec{x} \quad . \tag{1.15b}$$

Here $\vec{n}\times\vec{e}$ denotes the vector product. The explicit form of the solution (1.15) displays various properties:

(a) $\Phi(x)$ has exactly one zero: $\Phi(0) = 0$.

(b) $|\Phi(x)| < 1$.

(c) $1 - |\Phi(x)| = 1/r + (\tanh r - 1)/\tanh r$

$$= 1/r - 2e^{-2r} + O(e^{-3r}).$$

(d) $(A(x),\Phi(x))$ is real analytic on $\mathbb{R}^3$ (even though $\vec{n}$, r are undefined at $r = 0$).

These properties are quite similar to vortices, except the $1/r$ decay rate for $1 - |\Phi|$ which is special to $\lambda = 0$, $d = 3$. A computation shows

$$D_A\Phi = \mp \left(\frac{1}{r^2} - \frac{1}{\sinh^2 r} \right) (\vec{n}\cdot\vec{e})\vec{n}\cdot d\vec{x}$$

$$\mp \left(\frac{1}{\tanh r} - \frac{1}{r} \right)\left(\frac{1}{\sinh r} \right) (\vec{e} - (\vec{n}\cdot\vec{e})\vec{n})\cdot d\vec{x} \ . \qquad (1.16)$$

The two terms in (1.16) are the longitudinal and transverse components of $D_A\Phi$ with respect to Φ. In fact for Lie algebra valued functions χ, we define the decomposition

$$\chi = \chi_L + \chi_T \ ,$$
$$\chi_L = \Phi(\Phi,\chi)|\Phi|^{-2} \ . \qquad (1.17)$$

Clearly if $\chi(x)$ is smooth, then χ_T, χ_L are smooth on the complement of the (gauge invariant) set of zeros $Z(\Phi)$ of Φ,

$$Z(\Phi) = \{x: \Phi(x) = 0\} \ . \qquad (1.18)$$

Note

(e) $|(D_A\Phi)_L| \sim 1/r^2$.

(f) $|(D_A\Phi)_T| \sim e^{-r}$.

Thus we find exponential decay rates $m_L = 0$, $m_T = 1$ associated with the longitudinal and transverse modes. The properties (a) - (e) are generalized in the theorems below for finite action solutions to the second order equations (1.4)-(1.5) with $\lambda \geq 0$.

The first result is an existence theorem for the first order equations (1.14) with boundary conditions (1.5). Clearly we can restrict attention to $N > 0$, or to $N < 0$, for if $(A(x), \Phi(x))$ satisfies (1.14a), then $(A'(x), \Phi'(x)) = (A(x), -\Phi(x))$ satisfies (1.14b) and *vice versa.*

THEOREM 1.1. *Let* $N > 0$ *be a given integer. Let* $\{x_1, \ldots, x_N\} \in \mathbb{R}^3$ *be separated by a minimum distance* d,

$$d \equiv \min_{i \neq j} |x_i - x_j| \geq d_0 \quad . \tag{1.19}$$

There exists $d_0 < \infty$ *such that for any* $\{x_j\}$ *satisfying* (1.19), *there is a smooth solution* (A, Φ) *to* (1.14a), (1.5). *The solution is locally gauge equivalent to a real analytic solution. The monopole number is* $-N$ *and*

$$\frac{1}{4\pi} \operatorname{Tr} \int_{\mathbb{R}^3} (D_A \Phi) \wedge F = N \quad . \tag{1.20}$$

In addition,

$$0 \leq 1 - |\Phi| < \text{const} \cdot 1/r \quad , \tag{1.21}$$

$$|(D_A \Phi)_L| < \text{const} \cdot 1/r^2 \quad , \tag{1.22}$$

and given $\varepsilon > 0$,

$$|(D_A \Phi)_T| < \text{const}(\varepsilon) \cdot e^{-(1-\varepsilon) r} \quad . \tag{1.23}$$

The solution (A, Φ) of Theorem 1.1 can be regarded as an N-monopole solution, in the sense that one monopole is localized near each x_j, $j = 1, 2, \ldots, N$. Quantitatively we formulate this by showing that Φ has a zero near each x_j and that the local winding number of Φ, in the vicinity of each x_j, equals -1. Let

$$B_R(x_i) = \{x : |x - x_i| \le R\} \quad .$$

THEOREM 1.2. *Let* (A,Φ) *be a solution given by Theorem* 1.1. *There exist* $d_1 < \infty$ *and* $0 < r_0 \le 1/4$ *which do not depend on* (A,Φ) *such that if in addition to* (1.19),

$$R \equiv r_0 \left(\frac{d_1}{d}\right)^{1/2} < r_0 \quad , \tag{1.24}$$

then

(a) $Z(\Phi) \subset \bigcup_{i=1}^{N} B_R(x_i)$

(b) $Z(\Phi) \cap B_R(x_i) \neq \emptyset, \quad i = 1,\ldots,N.$

(c) $-\frac{1}{4\pi} \operatorname{Tr} \int_{\partial B_R(x_i)} (\hat{\Phi}\, d\hat{\Phi} \wedge d\hat{\Phi}) = 1, \quad i = 1,\ldots,N,$

where

$$\hat{\Phi} = \hat{\Phi}/|\Phi| \quad .$$

Remark. Condition (1.24) ensures that the balls $B_R(x_i)$ about each x_i do not overlap. Furthermore these balls may be chosen of radius $O(d^{-\frac{1}{2}})$ as $d \to \infty$. Then: (a) All zeros of Φ lie in these balls; (b) each ball contains some zero; and (c) the winding number on the boundary of each ball is one. Theorems 1.1-2 are proved in §8.

The first order equations (1.14) have many similarities to the equations of Chapter III. In that case we completely characterized the solutions by $Z(\Phi) = \{x_1,\ldots,x_N\}$. In the monopole case this appears insufficient; presumably $Z(\Phi)$ must be supplemented by other parameters. In fact E. Weinberg [22] has calculated that the dimension of the space of moduli is $4|N| - 1$. A physical interpretation of the extra $|N| - 1$ parameters is not known.

CONJECTURE 1. *The restriction* $|x_i - x_j| \geq d$ *can be removed from the theorem, and* N*-monopole solutions exist for arbitrary* $x_1, \ldots x_N$. *(See Rebbi and Rossi* [16], *for a numerical result for two monopoles superimposed. See also* [1,7] *for* O(2)*-symmetric ansatzes.)*

CONJECTURE 2. *The only finite action solutions to the second order equations* (1.4-5) *with* $\lambda = 0$ *are the monopole solutions to* (1.14a *or* b). *Hence we conjecture no monopole-antimonopole solutions exist.*

CONJECTURE 3. *Nontrivial critical points for* $\lambda > 0$ *and* $|N| \geq 2$ *exist. For* $|N| = 1$, *solutions are known* [17,21,6,13].

CONJECTURE 4. *Time dependent bound states of monopoles or monopole-antimonopole pairs exist.*

Next we give some results of a more general character which apply for all $0 \leq \lambda$, for the second order equations (1.4-5).

THEOREM 1.3. (Theorem V.1.1). *Let* (A,Φ) *be a finite action solution to* (1.4-5) *with* $0 \leq \lambda$. *Assume there exists a gauge in which the components of* (A,Φ) *and their first derivatives are locally square integrable. Then* (A,Φ) *is gauge equivalent to a globally* C^∞ *solution and locally gauge equivalent to a real analytic solution.*

With the justification of Theorem 1.3, we now restrict attention to smooth solutions (A,Φ) to the second order equations. We begin with the estimate which justifies the boundary conditions (1.5), as explained above.

THEOREM 1.4. *Let* (A,Φ) *be a smooth, finite action solution to* (1.4). *Then*

$$\lim_{R\to\infty} \inf_{|x|=R} |\Phi(x)| = c = \lim_{R\to\infty} \sup_{|x|=R} |\Phi(x)| < \infty \tag{1.25}$$

exists. Furthermore, either $\Phi(x) \equiv 0$, *or else for all* x,

$$|\Phi(x)| < c \quad . \tag{1.26}$$

If $0 < \lambda$, *then* $c = 1$.

THEOREM 1.5 (Theorem 10.5). *Let* (A,Φ) *be a smooth, finite action solution to* (1.4)-(1.5) *with* $\lambda = 0$. *Then* (1.20-23) *hold and* N *is an integer. In addition, for any* $\varepsilon > 0$,

$$|F_L(x)| \leq \text{const}/r^2 \quad , \tag{1.27}$$

$$|F_T(x)| \leq \text{const } e^{-(1-\varepsilon)r} \quad . \tag{1.28}$$

The constants depend on (A,Φ) *and on* ε.

THEOREM 1.6 (Theorem 10.2). *Let* (A,Φ) *be a smooth, finite action solution to* (1.4-5) *with* $0 < \lambda$. *Define* $m_L = \min(\lambda^{\frac{1}{2}},2)$ *and* $m_T = 1$. *Given* $\varepsilon > 0$,

$$|(D_A\Phi)_L(x)|, \qquad 1 - |\Phi(x)| \leq \text{const } e^{-(1-\varepsilon)m_L r} \quad , \tag{1.29}$$

$$|(D_A\Phi)_T(x)|, \qquad |F_T(x)| \leq \text{const } e^{-(1-\varepsilon)m_T r} \quad , \tag{1.30}$$

$$|F_L(x)| \leq \text{const}/r^2 \quad . \tag{1.31}$$

The constants depend on (A,Φ) *and on* ε.

In Sections 16 and 17 we return to the study of the solutions to (1.4) for the case $\lambda = 0$.

CONJECTURE 5. *If* (A,Φ) *is a finite action solution to* (1.4) *with* $\lambda = 0$, *then* $Z(\Phi)$ *is a finite set of points.*

This last conjecture is based on the parallels between equation (1.4) when $\lambda = 0$ and the vortex equation (III.1.4) when $\lambda = 1$ (cf. Theorems III.1.2-3). It follows from Theorems 1.3 and 1.6 that Φ

is real analytic and that $Z(\Phi)$ is a bounded set.

THEOREM 1.7. *Let* (A,Φ) *be a smooth, finite action solution to* (1.4) *with* $\lambda = 0$. *Then*

(a) $Z(\Phi)$ *has Lebesgue measure zero.*

(b) $Z(\Phi)$ *is not the boundary of an open set in* $\mathbb{R}^3$. (Proposition 16.2)

(c) *The maximum linear extent of a dimension* 1 *component of* $Z(\Phi)$ *is bounded by the action.* (Proposition 16.3)

(d) *Let* $S(\Phi) \subset \mathbb{R}^3$ *be the complement of the set on which* $\Phi/|\Phi|$ *is smooth* (Definition 16.4). *Then*

$$S(\Phi) = Z(\Phi) \quad .$$

We conclude for $\lambda = 0$ *that* Φ *is an order parameter. (This corresponds to the case in Chapter* III *with* $\lambda \leq 1$.)

THEOREM 1.8. (Theorems 17.1,2). *Let* (A,Φ) *be a smooth, finite action solution to the second order equations* (1.4-5) *with* $\lambda = 0$. *Then*

$$|F(x)| + |D_A\Phi(x)| \leq \mathrm{const}(1 - |\Phi(x)|) \qquad (1.32)$$

with a constant depending only on $\mathscr{A}(A,\Phi)$.

In the final sections of this chapter we state what is known about solutions to (1.4) for $\lambda = 0$ and gauge group $G \neq SU(2)$.

IV.2 THE FIRST ORDER EQUATIONS

The basic problem is to prove the existence of a solution to the first order monopole equation $*F = D_A\Phi$. Our method is to start from a chosen configuration (A_0,Φ_0) and look for a solution nearby. Define

$$a = A - A_0 , \quad \phi = \Phi - \Phi_0 , \tag{2.1}$$

with (a,ϕ) transforming under the adjoint representation of the gauge group. We show that the equations for (a,ϕ) have a solution, provided (A_0,ϕ_0) is sufficiently close to a solution to begin with. In particular, we require that each component

$$G_{0,i} \equiv -(*F_{A_0} - D_{A_0}\Phi_0)_i \tag{2.2}$$

have a small norm in certain L_p spaces. Constructing a solution is therefore reduced to finding a good starting point (A_0,Φ_0). We do this in §7 by patching together N copies of the one-monopole solution, taking widely separated single monopole configurations.

Let $a = \Sigma\, a_i dx^i$ and $(\nabla_{A_0})_j = \nabla_j + [(A_0)_j,\cdot]$. The equation $*F = D_A\Phi$, written in terms of the components of (a,ϕ) becomes

$$\left\{\varepsilon^{ijk}(\nabla_{A_0})_j a_k - (\nabla_{A_0})_i\phi + [\Phi_0,a_i]\right\} + [\phi,a_j] + \varepsilon^{ijk}a_j a_k = G_{0,i} . \tag{2.3}$$

The system (2.3) by itself is not elliptic, but it can be made elliptic by using the gauge freedom. We require that (a,ϕ) satisfy the background gauge condition

$$(\nabla_{A_0})_i a_i + [\Phi_0,\phi] = 0 . \tag{2.4}$$

Together (2.3-4) form an elliptic system. In order to deal with (2.3-4) as a simple unit, define unit quaternions $\{1,\tau_j : j = 1,2,3\}$ with

$$\tau_j\tau_k = -\delta_{jk} - \varepsilon_{ijk}\tau_k\ , \qquad \tau_k^* = -\tau_k \quad . \tag{2.5}$$

We consider the τ_j to be 2×2, anti-hermitian matrices which commute with the Lie algebra $\mathcal{g}$. Define a $\mathcal{g}$-valued quaternion ψ by

$$\psi = \phi + \tau_j a_j \quad .$$

Also let $\mathcal{D}_0$ denote the first order elliptic operator

$$\mathcal{D}_0 = -\tau_j(\nabla_{A_0})_j + [\Phi_0,\cdot] \quad . \tag{2.6}$$

Then the system (2.3-4) is equivalent to the system

$$\mathcal{D}_0\psi + \psi\wedge\psi = G_0 \quad , \tag{2.7}$$

when we define $\wedge$ to be a (symmetric) product on the $\mathcal{g}$-valued quaternion algebra. For $a = a_0 + a_j\tau_j$ and $b = b_0 + b_j\tau_j$,

$$a\wedge b = b\wedge a \equiv \frac{1}{2}\tau_i\left\{-[a_i,b_0] - [b_i,a_0] + \varepsilon^{ijk}(a_jb_k + b_ja_k)\right\} \quad . \tag{2.8}$$

Here

$$G_0 = -\tau_j(*F_{A_0} - D_{A_0}\Phi_0)_j = \tau_j G_{0,j} \quad ,$$

$$\psi\wedge\psi = -\tau_j[a_j,\phi] + \tau_i\varepsilon^{ijk}a_ja_k \quad . \tag{2.9}$$

Since $\det(\tau_j\xi_j) = |\xi|^2$, the symbol of $\mathcal{D}_0$ is uniformly elliptic. This justifies the statement that (2.7) is a properly elliptic first order system.

We introduce the standard scalar product on quaternion space Q. In particular consider the space $\mathcal{g}\otimes Q$ of $\mathcal{g}$-valued quaternions

$$a = a_0 + \tau_j a_j \quad . \tag{2.10}$$

For $a, b \in g \otimes Q$, define

$$(a,b) = \mathrm{Tr}(a_0^\dagger b_0 + a_j^\dagger b_j) \quad , \tag{2.11}$$

with Tr the normalized trace on g. Note that the inner product is gauge invariant.

To consider functions on $g \otimes Q$, we introduce the space C_0^∞ of smooth, compactly supported sections u of the vector bundle $\mathbb{R}^3 \times (g \otimes Q)$. We define a putative inner product by

$$\langle u,v\rangle_H = \langle\nabla_{A_0} u, \nabla_{A_0} v\rangle_{L_2} + \langle[\Phi_0,u],[\Phi_0,v]\rangle_{L_2} \quad , \tag{2.12}$$

where

$$\nabla_{A_0} u = \nabla_{A_0} u_0 + \tau_j \nabla_{A_0} u_j \quad ,$$

$$[\Phi_0,u] = [\Phi_0,u_0] + \tau_j[\Phi_0,u_j] \quad ,$$

and

$$\langle\ ,\ \rangle_{L_2} = \int_{\mathbb{R}^3} *(\ ,\ ) \quad .$$

Clearly

$$0 \leq \|u\|_H \equiv \langle u,u\rangle_H^{1/2} \quad .$$

In Chapter VI, Corollary 6.2, we prove the useful bound

$$\|u\|_{L_6} \leq c\,\|u\|_H \quad , \tag{2.13}$$

with a constant c independent of (A_0,Φ_0). It follows that $\|u\|_H = 0$ ensures $u = 0$, so $\|\cdot\|_H$ is a norm. Let $H = H(A_0,\Phi_0)$ denote the Hilbert space completion of C_0^∞ in this norm. One should bear in mind that the spaces $H(A_0,\Phi_0)$ in general depend on (A_0,Φ_0). The space H defined by our approximate solution (A_0,Φ_0) will be the space in which we solve (2.7). The basic existence theorem is:

THEOREM 2.1. (Theorem 3.1.) *Let* $(A_0,\Phi_0) \in C^\infty$ *have finite action and satisfy* (1.5). *There exists* $\varepsilon_0 > 0$ *sufficiently small, and independent of* (A_0,Φ_0), *such that if*

$$\left(1+\|\Phi_0\|_{L_\infty}\right)^2\left(\|G_0\|_{L_2}+\|G_0\|_{L_{6/5}}\right) \equiv \varepsilon < \varepsilon_0 \quad , \tag{2.14}$$

then there exists a solution $\psi \in H(A_0,\Phi_0)$ *to* (2.7). *Furthermore* ψ *is* C^∞ *and there exists a constant* $c < \infty$ *such that*

$$\|\psi\|_{L_2} \le c\varepsilon\left(1+\|\Phi_0\|_{L_\infty}\right)^{-2} \quad , \tag{2.15}$$

$$\|\psi\|_{H} \le c\varepsilon\left(1+\|\Phi_0\|_{L_\infty}\right)^{-1} \quad . \tag{2.16}$$

COROLLARY 2.2. (Corollary 3.7) $(A,\Phi) = (A_0 + a, \Phi_0 + \phi)$ *is a smooth, finite action solution to* $*F = D_A\Phi$.

The next result gives a pointwise bound on ψ. This allows us to conclude that if two configurations satisfy the conditions of Theorem 2.1 and the two (A_0,Φ_0) are sufficiently dissimilar, then the two solutions to $*F = D_A\Phi$ are distinct (and are not gauge equivalent). It is convenient to introduce the formal adjoint of $\mathcal{D}_0$,

$$\mathcal{D}_0^\dagger = -\tau_j(\nabla_{A_0})_j - [\Phi_0,\cdot] \quad . \tag{2.17}$$

THEOREM 2.3. (See §5.) *Let* (A_0,Φ_0) *satisfy the conditions of Theorem* 2.1. *Suppose* $\mathcal{D}_0^\dagger G_0 \in L_2$. *Then there is a constant* c_2 *independent of* (A_0,Φ_0) *such that*

$$\|\psi\|_{L_\infty} \le c_2\left[\|\mathcal{D}_0^\dagger G_0\|_{L_2} + \|\psi\|_{L_2} + (\mathcal{A}(A_0,\Phi_0) + \|\psi\|_H^2)(\|\psi\|_H + \|\psi\|_{L_2})\right] . \tag{2.18}$$

Finally we give a result special to $G = SU(2)$, which says that (A,Φ) and (A_0,Φ_0) have the same winding number.

DEFINITION 2.4. *Let* $G = SU(2)$. *The set* $\mathscr{H}_N$ *is the set of* $(A,\Phi) \in C^\infty$ *and satisfying*

(a) $\mathscr{A}(A,\Phi) < \infty$.

(b) *Uniform boundary condition* (1.5).

(c) *Monopole number* $N = \frac{1}{4\pi} \operatorname{Tr} \int_{S^2_\infty} \Phi\, F$.

THEOREM 2.5. (See §5). *Let* $(A_0,\Phi_0) \in \mathscr{H}_N$, *and suppose* $(A_0 + a, \Phi_0 + \phi)$ *is a finite action solution to* (1.4). *Suppose further that* $\psi = \phi + \tau_j a_j \in L_2 \cap H$. *Then* $(A_0 + a, \Phi_0 + \phi) \in \mathscr{H}_N$ *also.*

IV.3. L_p THRESHOLD FOR MONOPOLE EXISTENCE

The idea to establish a threshold for monopole existence is to write $\psi = \mathscr{D}_0^\dagger u$, and to replace (2.7) by the equation

$$\mathscr{D}_0 \mathscr{D}_0^\dagger u + \mathscr{D}_0^\dagger u \wedge \mathscr{D}_0^\dagger u = G_0 \quad . \tag{3.1}$$

Essentially, we establish an implicit function theorem to invert (3.1). We find sufficient conditions on G_0 to ensure the existence of a regular solution u, yielding $\mathscr{D}_0^\dagger u$ which satisfies (2.7). Throughout this section

$$\Phi_0 : \mathbb{R}^3 \to \mathfrak{g} \,, \qquad A_0 : \mathbb{R}^3 \to T^* \otimes \mathfrak{g} \quad , \tag{3.2}$$

$$u : \mathbb{R}^3 \to \mathfrak{g} \otimes \mathbb{Q} \quad .$$

The solution u will be given by a convergent expansion

$$u = \sum_{n=0}^{\infty} u_n \quad , \tag{3.3}$$

where the expansion parameter is

$$\varepsilon \equiv \left(1 + \|\Phi_0\|_{L_\infty}\right)^2 \left(\|G_0\|_{L_2} + \|G_0\|_{L_{6/5}}\right) \quad . \tag{3.4}$$

THEOREM 3.1. *Let* $(A_0, \Phi_0) \in C^\infty$ *and satisfy* (1.5). *There exists* $\varepsilon_0 > 0$ *and* $c < \infty$, *independent of* (A_0, Φ_0), *such that if*

$$\varepsilon < \varepsilon_0 \quad , \tag{3.5}$$

then there exists a C_∞ *solution* u *to* (3.1) *satisfying*

$$\|u\|_H \leq \varepsilon c \left(1 + \|\Phi_0\|_{L_\infty}\right)^{-2} \quad , \tag{3.6}$$

$$\|\mathscr{D}_0^\dagger u\|_H \leq \varepsilon c \left(1 + \|\Phi_0\|_{L_\infty}\right)^{-1} \quad . \tag{3.7}$$

Furthermore $\psi = \mathscr{D}_0^\dagger u$ *satisfies* (2.7).

COROLLARY 3.2. $\psi \in L_2 \cap H$.

Proof. From (3.7), we conclude that $\psi = \mathscr{D}_0^\dagger u \in H$. Furthermore, $\psi \in L_2$, since

$$\|\psi\|_{L_2} = \|\mathscr{D}_0^\dagger u\|_{L_2} = \|-\tau_k(\nabla_{A_0})_k u - [\Phi_0, u]\|_{L_2} \leq 2\|u\|_H \ ,$$

by the triangle inequality. The desired bound results from (3.6).

In order to study the nonlinear equation (3.1) we use properties of the linear equation

$$\mathscr{D}_0 \mathscr{D}_0^\dagger u = q \quad , \tag{3.8}$$

as an equation for $u \in H(A_0, \Phi_0)$. The following theorem is proved in §4.

THEOREM 3.3. *Let* $(A_0,\Phi_0) \in C^\infty$ *and satisfy* (1.5). *Let* $q \in L_2 \cap L_{6/5} \cap C^\infty$. *Let* $\|G_0\|_{L_{3/2}}$ *be sufficiently small,*

$$\|G_0\|_{L_{3/2}} < \varepsilon_1 \quad , \tag{3.9}$$

where $0 < \varepsilon_1$ *is a constant independent of* (A_0,Φ_0,q). *Then there exists a* C^∞ *solution* $u \in H$ *to* (3.8) *such that* $\mathscr{D}_0^\dagger u \in H$, *and*

$$\|u\|_H \le c_1 \|q\|_{L_{6/5}} \quad , \tag{3.10}$$

$$\|\mathscr{D}_0^\dagger u\|_H \le c_1 \left(\|q\|_{L_2} + \|\Phi_0\|_{L_\infty} \|q\|_{L_{6/5}} \right) \quad , \tag{3.11}$$

and for $2 \le p \le 6$,

$$\|\mathscr{D}_0^\dagger u\|_{L_p} \le c_1 \left(\|q\|_{L_2} + \left(1 + \|\Phi_0\|_{L_\infty}\right) \|q\|_{L_{6/5}} \right) \quad . \tag{3.12}$$

Here $c_1 < \infty$ *is a constant independent of* (A_0,Φ_0,q). *Also* u *is unique.*

<u>Proof of Theorem 3.1, Assuming Theorem 3.3.</u> Let us begin with the formal aspects of the proof. We define u_k recursively as the solution to the linear equation

$$\mathscr{D}_0 \mathscr{D}_0^\dagger u_k = q_k \tag{3.13}$$

where

$$q_0 = G_0, \qquad q_1 = -\mathscr{D}_0^\dagger u_0 \wedge \mathscr{D}_0^\dagger u_0 \tag{3.14}$$

and for $k \ge 2$,

$$q_k = -\sum_{j=0}^{k-2} (\mathscr{D}_0^\dagger u_j \wedge \mathscr{D}_0^\dagger u_{k-1} + \mathscr{D}_0^\dagger u_{k-1} \wedge \mathscr{D}_0^\dagger u_j) - \mathscr{D}_0^\dagger u_{k-1} \wedge \mathscr{D}_0^\dagger u_{k-1} \; . \tag{3.15a}$$

Assuming the solutions u_k exist, they satisfy the identity

$$\sum_{k=0}^{m} q_k = G_0 - \sum_{k,\ell=0}^{m-1} \mathscr{D}_0^\dagger u_k \wedge \mathscr{D}_0^\dagger u_\ell \quad . \tag{3.15b}$$

Define the partial sum

$$S_m \equiv \sum_{k=0}^{m} u_k \quad .$$

As a consequence of (3.15),

$$\mathscr{D}_0 \mathscr{D}_0^\dagger S_m = G_0 - \mathscr{D}_0^\dagger S_{m-1} \wedge \mathscr{D}_0^\dagger S_{m-1} \quad . \tag{3.16}$$

Hence if $\{S_m\}$ has a limit u, then u is the desired solution. We now prove that each u_k does exist, that $u_k \in C^\infty \cap H(A_0,\Phi_0)$ and that S_m converges in $H(A_0,\Phi_0)$ to a solution u to (3.1).

The proof is inductive. At each stage of the induction we require that G_0 be small. This allows us to use Theorem 3.3 to invert $\mathscr{D}_0 \mathscr{D}_0^\dagger$ as a linear operator; the first step is to solve

$$\mathscr{D}_0 \mathscr{D}_0^\dagger u_0 = G_0 \quad . \tag{3.17}$$

We again use the fact that the inhomogeneous term in (3.17) has small norm to conclude that u_0 has small norm.

To begin, note that (3.4) ensures $\|G_0\|_{L_2} < \varepsilon$, $\|G_0\|_{L_{6/5}} < \varepsilon$. But $6/5 < 3/2 < 2$. Therefore there is a constant such that

$$\|G_0\|_{L_{3/2}} < \text{const}\ \varepsilon \quad , \tag{3.18}$$

and, for ε sufficiently small, (3.9) holds. By Theorem 3.3, a C^∞ solution to (3.17) exists, and using (3.4),

$$\|u_0\|_H \le c_1 \left(1 + \|\Phi_0\|_{L_\infty}\right)^{-2} \varepsilon \quad , \tag{3.19}$$

$$\|D_0^\dagger u_0\|_H \le c_1 \left(1 + \|\Phi_0\|_{L_\infty}\right)^{-1} \varepsilon \quad . \tag{3.20}$$

We now obtain similar results for u_k.

LEMMA 3.4. *Let* ε_0 *in* (3.5) *be sufficiently small. Each* u_k, q_k *exists and is* C^∞. *There is a constant* $b < \infty$ *independent of* (A_0, Φ_0) *such that for each positive integer* k,

$$\|q_k\|_{L_2} + \|q_k\|_{L_{6/5}} \le b^k \varepsilon^{k+1} \left(1 + \|\Phi_0\|_{L_\infty}\right)^{-2} \quad . \tag{3.21}$$

Proof. We note that it is sufficient to establish (3.21). Having done this, Theorem 3.3 ensures the existence of a C^∞ solution u_k to (3.13), and hence allows the construction of q_{k+1}, u_{k+1}. For $k = 0$, the bound (3.4) is exactly (3.21). Also for $k = 1$, by (3.12), (3.4) and for $1 \le p \le 3$,

$$\frac{1}{2} \|q_1\|_{L_p} \le \|\mathscr{D}_0^\dagger u_0\|_{L_{2p}}^2 \le c_1^2 \varepsilon^2 \left(1 + \|\Phi_0\|_{L_\infty}\right)^{-2}$$

so

$$\|q_1\|_{L_2} + \|q_1\|_{L_{6/5}} \le b\varepsilon^2 \left(1 + \|\Phi_0\|_{L_\infty}\right)^{-2} \quad , \tag{3.22}$$

when $b \ge 2c_1^2$.

We now assume (3.4), (3.22) and (3.21) for $k = 2,3,\ldots,r-1$ and establish it for $k = r$. In fact

$$|q_k| \le 4 \sum_{j=0}^{k-1} |\mathscr{D}_0^\dagger u_j| \, |\mathscr{D}_0^\dagger u_{k-1}|$$

so by Hölder's inequality, and for $r \ge 2$, $1 \le p \le 3$,

$$\|q_r\|_{L_p} \le 4 \sum_{j=0}^{r-1} \|\mathscr{D}_0^{\dagger} u_j\|_{L_{2p}} \|\mathscr{D}_0^{\dagger} u_{r-1}\|_{L_{2p}}$$

$$\le 4c_1^2\left(1+\|\Phi_0\|_{L_\infty}\right)^2 \left(\|q_{r-1}\|_{L_2} + \|q_{r-1}\|_{L_{6/5}}\right) \sum_{j=0}^{r-1} \left(\|q_j\|_{L_2} + \|q_j\|_{L_{6/5}}\right)$$

$$\le 4c_1^2 \left(1+\|\Phi_0\|_{L_\infty}\right)^{-2} b^{r-1}\varepsilon^{r+1} \sum_{j=0}^{r-1} (b\varepsilon)^j$$

$$\le \left(4c_1^2 \sum_{j=0}^{\infty} (b\varepsilon)^j b^{-1}\right) b^r \varepsilon^{r+1} \left(1 + \|\Phi_0\|_{L_\infty}\right)^{-2}$$

$$\le \frac{1}{2} b^r \varepsilon^{r+1} \left(1 + \|\Phi_0\|_{L_\infty}\right)^{-2} \quad . \tag{3.23}$$

In the last step we choose $b \ge 16c_1^2$ and $\varepsilon_0 \le (2b)^{-1}$. Since $\varepsilon < \varepsilon_0$ and for $z < 1$, $\Sigma_{k=1}^{\infty} z^k = 1/1-z$,

$$2c_1^2(1 - b\varepsilon)^{-1} b^{-1} \le \frac{1}{2} \quad .$$

The inequality (3.21) follows.

LEMMA 3.5. *The sequences* S_m *and* $D_0^{\dagger} S_m$ *converge in* $H(A_0, \Phi_0)$ *to limits* u *and* ψ. *Furthermore*

$$\psi = \mathscr{D}_0^{\dagger} u \quad . \tag{3.24}$$

<u>Proof</u>. From Theorem 3.3 we obtain estimates on $\|u_k\|_H$ and $\|D_0^{\dagger} u_k\|_H$. In particular

$$\|u_k\|_H \le c_1 \|q_k\|_{L_{6/5}} \le c_1 \left(1 + \|\Phi_0\|_{L_\infty}\right)^{-2} b^k \varepsilon^{k+1} \tag{3.25}$$

and

$$\|D_0^{\dagger}u_k\|_H \le c_1\left(1+\|\Phi_0\|_{L_\infty}\right)\left(\|q_k\|_{L_2}+\|q_k\|_{L_{6/5}}\right)$$

$$c_1\left(1+\|\Phi_0\|_{L_\infty}\right)^{-1} b^k\varepsilon^{k+1} \qquad . \tag{3.26}$$

Therefore for $m \ge n$, we estimate the tail of the series for S_m,

$$\|S_m - S_n\|_H \le c_1\left(1+\|\Phi_0\|_{L_\infty}\right)^{-2}(1-b\varepsilon_0)^{-1}b^n\varepsilon^{n+1} \quad ,$$

and

$$\|D_0^{\dagger}(S_m - S_n)\|_H \le c_1\left(1+\|\Phi_0\|_{L_\infty}\right)^{-1}(1-b\varepsilon_0)^{-1}b^n\varepsilon^{n+1} \quad .$$

Thus $\{S_m\}$ and $\{D_0^{\dagger}S_m\}$ converge in H. Furthermore, $u = \lim S_m$ is in the domain of the (closed) operator $\mathscr{D}_0^{\dagger}$ and $\mathscr{D}_0 u = \psi$.

LEMMA 3.6. *The functions*

$$v_m = \mathscr{D}_0\mathscr{D}_0^{\dagger}S_m + \mathscr{D}_0^{\dagger}S_m \wedge \mathscr{D}_0^{\dagger}S_m - G_0$$

converge to zero in L_2.

Proof. Using the identity (3.16), and (3.26),

$$\|v_{m-1}\|_{L_2} = \|\mathscr{D}_0\mathscr{D}_0^{\dagger}u_m\|_{L_2} \le \|\tau_k(\nabla_{A_0})_k\mathscr{D}_0^{\dagger}u_m\|_{L_2} + \|[\Phi_0,\mathscr{D}_0^{\dagger}u_m]\|_{L_2}$$

$$\le 2\|\mathscr{D}_0^{\dagger}u_m\|_H \le O(\varepsilon^{m+1}) \to 0 \qquad .$$

Proof of Theorem 3.1 (completion). Since $v_m \to 0$ in L_2, $u = \lim S_m$ is a weak solution to (3.1). The claim that $u \in C^\infty$ is a standard inductive argument on the greatest integer k such that $u \in L_1^k$, locally. See Chapter VI, Section 9, and also [12], Chapter 6. The bounds (3.6-3.7) follow by summing (3.25-3.26). For ε_0, b as above, we may choose $c = 2c_1$, and the theorem follows.

COROLLARY 3.7. *The solution* $(A,\Phi) = (A_0 + a, \Phi_0 + \phi)$ *is a* C^∞, *finite action solution to* $*F = D_A\Phi$.

Proof. Since u is C^∞, and (A_0,Φ_0) is C^∞, thus a, ϕ is C^∞, and hence also (A,Φ). That (A,Φ) has finite action is a consequence of the finite action of (A_0,Φ_0) and Theorem 3.1. In fact,

$$F_A = F_{A_0} + D_{A_0} a + a \wedge a \quad , \tag{3.27}$$

$$D_A\Phi = D_{A_0}\Phi_0 + D_{A_0}\phi + [a,\phi] + [a,\Phi_0] \quad . \tag{3.28}$$

We appeal to the inequality (2.13), namely

$$\|u\|_{L_6} \le \text{const } \|u\|_H \quad , \tag{3.29}$$

to argue that each term in (3.27-28) is L_2.

Since (A_0,Φ_0) has finite action, $F_{A_0} \in L_2$. Furthermore $\psi \in H$, so $D_{A_0} a \in L_2$. The last term in (3.27) is estimated by

$$\begin{aligned} \|a \wedge a\|_{L_2} &\le \text{const } \|a\|_{L_4}^2 \le \text{const } \|\psi\|_{L_4}^2 \\ &\le \text{const } \left(\|\psi\|_{L_2} + \|\psi\|_{L_6}^3 \right) \\ &\le \text{const } \left(\|\psi\|_{L_2} + \|\psi\|_H^3 \right) \end{aligned} \tag{3.30}$$

Since $\psi \in L_2 \cap H$, we conclude that $F_A \in L_2$.

Now inspect (3.28). The condition that (A_0,Φ_0) has finite action ensures $D_{A_0}\Phi_0 \in L_2$. Furthermore $\psi \in H$ ensures $D_{A_0}\phi \in L_2$. The term $[a,\phi]$ in (3.28) is estimated as in (3.30), using

$$\|[a,\phi]\|_{L_2} \leq \text{const}\,\|a\|_{L_4}\,\|\phi\|_{L_4} \leq \text{const}\,\|\psi\|_{L_4}^2 \quad .$$

Finally

$$\|[a,\Phi_0]\|_{L_2} \leq \|\psi\|_H \quad ,$$

so each term in (3.28) is also L_2.

IV.4. THE LINEAR EQUATION $\mathscr{D}_0\,\mathscr{D}_0^\dagger u = q$

In this section we study the equation $\mathscr{D}_0\mathscr{D}_0^\dagger u = q$ and prove Theorem 3.3. We begin by remarking that for $\mathscr{D}_0, \mathscr{D}_0^\dagger$ defined in (2.6), (2.17),

$$\mathscr{D}_0\,\mathscr{D}_0^\dagger u = -\nabla_{A_0}^2 u - [\Phi_0,[\Phi_0,u]] + \tau_j[G_{0,j},u] \quad . \tag{4.1}$$

The cyclic property of the trace on $\mathfrak{g}$ implies that for any pair of elements $\alpha,\ \beta \in \mathfrak{g} \otimes \mathbb{Q}$,

$$(\alpha,\tau_j[G_{0,j},\beta]) = (\tau_j[G_{0,j},\alpha],\beta) \quad . \tag{4.2}$$

If the components of G_0 are L_{2p}, $1 \leq p \leq \infty$, then $\tau_j[G_{0,j},\cdot]$ extends to a hermitian operator on L_2 with domain $L_{2p/(p-1)}$, by Hölder's inequality. Thus the solution to $\mathscr{D}_0\mathscr{D}_0^\dagger u = q$ is formally the critical point of the real functional

$$\begin{aligned} S[u] &= \tfrac{1}{2}\langle\nabla_{A_0}u,\nabla_{A_0}u\rangle_{L_2} + \tfrac{1}{2}\langle[\Phi_0,u],[\Phi_0,u]\rangle_{L_2} \\ &\quad + \tfrac{1}{2}\langle u,\tau_j[G_{0,j},u]\rangle_{L_2} - \langle u,q\rangle_{L_2} \\ &= \tfrac{1}{2}\|\mathscr{D}_0^\dagger u\|_{L_2}^2 - \langle u,q\rangle_{L_2} \quad . \end{aligned} \tag{4.3}$$

We now study the quadratic form $S[u]$ for $u \in H$. Before continuing the analysis of $S[u]$, we mention that as before

$$\Phi_0 : \mathbb{R}^3 \to g, \qquad A_0 : \mathbb{R}^3 \to T^* \otimes g, \qquad q : \mathbb{R}^3 \to g \otimes Q, \quad \text{etc.}$$

PROPOSITION 4.1. *Let* $(A_0, \Phi_0) \in C^\infty$ *satisfy* (1.5). *There exists a constant* $\varepsilon_1 > 0$, *independent of* (A_0, Φ_0), *with the following property. Suppose* $\|G_0\|_{L_{3/2}} < \varepsilon_1$ *and* $q \in L_{6/5}$. *Then there is a unique* $u \in H$ *which is a weak solution to* (4.4b) *in the sense that for all* $v \in H$,

$$\langle \nabla_{A_0} v, \nabla_{A_0} u \rangle_{L_2} + \langle [\Phi_0, v], [\Phi_0, u] \rangle_{L_2} + \langle v, \tau_k [G_{0,k}, u] \rangle_{L_2} - \langle v, q \rangle_{L_2} = 0. \tag{4.4a}$$

If $q \in C^\infty$, *then also* $u \in C^\infty$ *and*

$$\mathcal{D}_0 \mathcal{D}_0^\dagger u = q \quad . \tag{4.4b}$$

Proof. We show that $S[u]$ has a unique minimum on H for $\|G_0\|_{L_{3/2}}$ sufficiently small. We use the calculus of variations. (See Chapter VI.8.) In particular we show that $S[u]$ is defined on H, is strictly convex, differentiable and satisfies a coercive bound.

LEMMA 4.3. *The functional* $S[u]$ *is continuous on* H *and satisfies a coercive lower bound. In fact for* ε_1 sufficiently small,

$$-c_1 + \frac{1}{4} \|u\|_H^2 \leq S[u] \leq c_2 \|u\|_H^2 + c_1 \quad , \tag{4.5}$$

where c_1 *depends only on* $\|q\|_{L_{6/5}}$ *and* c_2 *depends only on* $\|G_0\|_{L_{3/2}}$.

Proof. The upper bound follows from Hölder's inequality. In fact

$$|S[u]| \leq \frac{1}{2} \|u\|_H^2 + 2 \| |u|^2 |G_0| \|_{L_1} + |\langle u, q \rangle_{L_2}|$$

$$\leq \frac{1}{2} \|u\|_H^2 + 2 \|u\|_{L_6}^2 \|G_0\|_{L_{3/2}} + \|u\|_{L_6} \|q\|_{L_{6/5}} \quad .$$

Using (2.13), and $\|u\|_H \le \delta\|u\|_H^2 + \delta^{-1}$, we have (4.5). Similarly, the lower bound follows from

$$S[u] \ge \frac{1}{2}\|u\|_H^2 - 2c\,\|u\|_H^2\,\|G_0\|_{L_{3/2}} - c\,\|u\|_H\,\|q\|_{L_{6/5}}$$

which yields (4.5) when $\|G_0\|_{L_{3/2}}$ is sufficiently small.

LEMMA 4.4. *The functional* $S[u]$ *is differentiable on* H *and*

$$\operatorname{grad} S[u;v] \equiv D_v S[u] = \langle v,u\rangle_H + \langle v,\tau_k[G_{0,k},u]\rangle_{L_2} - \langle v,q\rangle_{L_2}, \quad (4.6)$$

is jointly continuous in v, $u \in H$.

Proof. Let $S_q[u]$ denote the q-dependence of S. The difference quotient for the directional derivative is

$$t^{-1}(S_q[u+tv] - S_q[u]) = \mathscr{L} + tS_0[v] \quad ,$$

where $\mathscr{L}$ denotes (4.6). By (4.5), $S_0[v]$ is defined on H and hence the difference quotient converges to $\mathscr{L}$. Furthermore, Hölder's inequality applied in a similar manner as in the previous lemma shows joint H continuity of $\mathscr{L}$ in u and v.

LEMMA 4.5. *The norm* $\|u\|_H$ *is equivalent to the norm* $\|\mathscr{D}_0^{\dagger}u\|_{L_2} = (2S_0[u])^{\frac{1}{2}}$.

Proof. With $q=0$, $S[u] = \frac{1}{2}\|\mathscr{D}_0^{\dagger}u\|_{L_2}^2$. But in this case we may take the constant $c_1 = 0$ in (4.5).

LEMMA 4.6. *The functional* $S[u]$ *is strictly convex.*

Proof. Since $\langle u,q\rangle_{L_2}$ is linear and continuous, it is convex. Hence it is sufficient to prove that $S_0[u]$ is strictly convex. Since $u \to \|\mathscr{D}_0^{\dagger}u\|_{L_2}$ is an equivalent norm for H, it cannot vanish unless

$u = 0$. Thus Chapter VI, Proposition 7.9 ensures strict convexity of $S_0[u] = \frac{1}{2} \|\mathscr{D}_0^\dagger u\|_{L_2}^2$.

Proof of Proposition 4.1 (completion). Because $S[u]$ is differentiable and convex, it is weakly lower semicontinuous by Proposition VI.7.8. It is coercive and strictly convex, so by Proposition VI.8.5 it has a unique critical point $\bar{u} \in H$, and $\bar{u}$ minimizes $S[u]$. Thus (4.4) holds, i.e. $\text{grad}\, S[\bar{u}; \cdot] = 0$. For $q \in C^\infty$, standard elliptic regularity ensures $u \in C^\infty$ [12].

We now prove Theorem 3.3. The existence of a solution follows from Proposition 4.1. We still require the *a priori* bounds (3.10-12)

PROPOSITION 4.7. *Let* $(A_0, \Phi_0) \in C^\infty$ *satisfy conditions of Proposition* 4.1. *Suppose in addition* $q \in L_2$, *and let* u *be the unique weak solution to* (4.4). *Then* $\mathscr{D}_0^\dagger u \in H$ *and*

$$\|u\|_H \le c\, \|q\|_{L_{6/5}} \quad , \tag{4.7}$$

$$\|\mathscr{D}_0^\dagger u\|_H \le c \left(\|q\|_{L_2} + \|\Phi_0\|_{L_\infty} \|q\|_{L_{6/5}} \right) , \tag{4.8}$$

and for $2 \le p \le 6$,

$$\|\mathscr{D}_0^\dagger u\|_{L_p} \le c \left(\|q\|_{L_2} + \|\Phi_0\|_{L_\infty} \|q\|_{L_{6/5}} \right) , \tag{4.9}$$

where c *is a constant independent of* (A_0, Φ_0, q).

LEMMA 4.8. *There is a constant* $c < \infty$, *independent of* (A_0, Φ_0) *such that for all* $v \in H \cap L_2$,

$$\|v\|_H \le c \left(\|\mathscr{D}_0 v\|_{L_2} + \|\Phi_0\|_{L_\infty} \|v\|_{L_2} \right) \quad . \tag{4.10}$$

Proof. For $v \in C^\infty$, $\mathscr{D}_0^\dagger v = \mathscr{D}_0 v - 2[\Phi_0, v]$. Thus by Lemma 4.5,

$$\|v\|_H \le \text{const}\|\mathscr{D}_0^\dagger v\|_{L_2} \le \text{const}\left(\|\mathscr{D}_0 v\|_{L_2} + 2\,\|\Phi_0\|_{L_\infty}\,\|v\|_{L_2}\right). \tag{4.11}$$

Likewise

$$\|\mathscr{D}_0 v\|_{L_2} \le \|\mathscr{D}_0^\dagger v\|_{L_2} + 2\|[\Phi_0,v]\|_{L_2} \le \text{const}\left(\|v\|_H + \|\Phi_0\|_{L_\infty}\,\|v\|_{L_2}\right).$$

Hence (4.11) extends by continuity to $v \in H \cap L_2$.

LEMMA 4.9. *Let* $v \in L_p$ *for some* $2 \le p \le 6$, *and suppose that* $\|v\|_H < \infty$. *Then* $v \in H$.

Proof. Assume first that $v \in C^\infty$. The space H is by definition the closure of C_0^∞ in the $\|\cdot\|_H$ norm. For $b_R(x)$ a C_0^∞ function, $b_R v \in H$. We choose as b_R the function VI(3.6) and show that $b_R v$ is Cauchy in H as $R \to \infty$ with limit v. Since $\|v\|_H < \infty$, it follows that $[\Phi_0,v]$ and $\nabla_{A_0}v$ have finite L_2 norm. Hence they are L_2 functions. (In fact every f with $\|f\|_{k,p} < \infty$ is an L_p^k function; [VI.1], page 52.) Thus $b_R[\Phi_0,v]$, $b_R\nabla_{A_0}v$ converge strongly to $[\Phi_0,v]$, $\nabla_{A_0}v$, respectively, in L_2 and $(1-b_R)v$ converges strongly to zero in L_p (see Proposition VI.1.3). But

$$\nabla_{A_0}(b_R v) = b_R\nabla_{A_0}v + (\nabla b_R)v$$

and

$$\|(\nabla b_R)v\|_{L_2} = \|(\nabla b_R)(1-b_{R/2})v\|_{L_2} \le \text{const}\cdot R^{\frac{p-6}{2p}}\,\|(1-b_{R/2})v\|_{L_p}.$$

Hence $b_R v \to v$ and $v \in H$.

Next assume $v \notin C^\infty$. Since $b_R v$ is compactly supported and A_0, Φ_0 are smooth, $b_R v \in L_2^1$; in fact for all w supported in a fixed compact set in B_{2R},

$$\|w\|_{1,2} \le \text{const}_R\left(\|\nabla_A w\|_{L_2} + \|w\|_{L_6}\right) \le \text{const}_R\,\|w\|_H.$$

Thus there exists a sequence $\{v_{R,n}\}$ in C_0^∞, supported in the ball B_{2R}, such that $v_{R,n} \to b_R v$ in L_2^1. Also for all w supported in B_{2R},

$$\|w\|_H \leq \text{const}_R \|w\|_{1,2} \quad .$$

Thus $\{v_{R,n}\}$ converges in H also, and

$$v_{R,n} \xrightarrow[H]{} b_R v \quad .$$

Thus $b_R v \in H$. By the previous argument $b_R v \to v$ in H, so $v \in H$ as well.

Proof of Proposition 4.7. Since (4.4) holds for every v, choose $u = v$. As in the proof of Lemma 4.3, we have by Kato's and Hölder's inequalities

$$2S_0[u] = \langle u,q \rangle_{L_2} \leq c \|u\|_H \|q\|_{L_{6/5}} \quad .$$

Using Lemma 4.5, and making a new choice of c, (4.7) follows.

We use Lemma 4.8 with $v = \mathscr{D}_0^\dagger u$. If $q \in C^\infty$, then $u \in C^\infty$ and since $\mathscr{D}_0 \mathscr{D}_0^\dagger u = q$,

$$\| \mathscr{D}_0^\dagger u\|_H \leq c \left(\|q\|_{L_2} + \|\Phi_0\|_{L_\infty} \|\mathscr{D}_0^\dagger u\|_{L_2} \right) \quad .$$

Again using Lemma 4.5 and estimate (4.7), the inequality (4.8) follows. By Lemma 4.9, $\mathscr{D}_0^\dagger u \in H$.

By Kato's inequality and (4.8)

$$\| \mathscr{D}_0^\dagger u\|_{L_6} \leq \|\mathscr{D}_0^\dagger u\|_H \leq c \left(\|q\|_{L_2} + \|\Phi\|_{L_\infty} \|q\|_{L_{6/5}} \right) \quad .$$

Also by Lemma 4.5, and (4.7),

$$\| \mathscr{D}_0^\dagger u\|_{L_2} \le \text{const}\ \|u\|_H \le c\ \|q\|_{L_{6/5}} \quad .$$

Hence $\| \mathscr{D}_0^\dagger u\|_{L_p}$ satisfies (4.9).

IV.5. HOW CLOSE IS (A_0,Φ_0)?

Suppose $(A,\Phi) = (A_0+a,\Phi_0+\phi)$ is a monopole solution to (1.14). The methods of §2-3 yield such a solution, as long as $*F_{A_0} - D_{A_0}\Phi_0$ is small in certain L_p norms. In this section we derive bounds on $|a(x)|$, $|\phi(x)|$ in terms of these norms and of the action $\mathscr{A}(A_0,\Phi_0)$. Because $|a(x)|$, $|\phi(x)|$ are gauge invariant, they give a point by point statement of how close (A_0,Φ_0) is to the solution. Again, for convenience, we use $\psi(x) = \phi(x) + \tau_j a_j(x)$, and estimate

$$\psi(x) = (|\phi(x)|^2 + |a(x)|^2)^{1/2} \quad .$$

We also show here that for $G = SU(2)$, the monopole numbers for (A,Φ) and for (A_0,Φ_0) are the same. The precise statements of these results are in Theorems 2.3 and 2.6, which we now prove.

Proof of Theorem 2.3. Let $\psi \in C^\infty$ be the solution to equation (2.7) given by Theorem 2.1. Differentiating yields

$$\mathscr{D}_0^\dagger \mathscr{D}_0\psi = -\mathscr{D}_0^\dagger(\psi\wedge\psi) + \mathscr{D}_0^\dagger G_0 \quad . \tag{5.1}$$

Using the definitions of $\mathscr{D}_0^\dagger, \mathscr{D}_0$ we have,

$$-\nabla_{A_0}^2\psi - [\Phi_0,[\Phi_0,\psi]] - \tau_k[(*F_{A_0} + D_{A_0}\Phi_0)_k,\psi] = \mathscr{D}_0^\dagger G_0 - \mathscr{D}_0^\dagger(\psi\wedge\psi) . \tag{5.2}$$

We remark that (5.2) is of the type considered in Proposition V.7.2, since

$$-(\psi,\nabla_A^2\psi) \le |\bar{G}_0||\psi|^2 + (\psi,\mathscr{D}_0^\dagger G_0) + 4|\nabla_{A_0}\psi||\psi|^2 + 4|[\Phi_0,\psi]||\psi|^2. \tag{5.3}$$

where $\bar{G}_0 = \tau_k \bar{G}_{0,k} = \tau_k(*F_{A_0} + D_{A_0}\Phi_0)_k$. In the notation of Chapter V, set

$$|\Lambda| = |\bar{G}_0| + 4(|\nabla_{A_0}\psi| + |[\Phi_0,\psi]|) \quad ,$$

$$|v| = |\mathscr{D}_0^\dagger G_0| \quad . \tag{5.4}$$

Then from Proposition V.7.2, we conclude that

$$\|\psi\|_\infty \le c_2 \left(\|\psi\|_{L_2} + \|\mathscr{D}_0^\dagger G_0\|_{L_2} + 2\left(\|\nabla_{A_0}\psi\|_{L_2} + \|\psi\|_{L_2}\right)\left(\|G_0\|_{L_2}^2 + 8\|\psi\|_H^2\right)\right) \tag{5.5}$$

with $c_2 < \infty$, independent of (A_0,Φ_0). Equation (2.18) follows by substituting

$$\|\bar{G}_0\|_{L_2}^2 \le 4\mathscr{A}(A_0,\Phi_0) \quad . \tag{5.6}$$

Proof of Theorem 2.5 assuming Theorem 10.5. It follows from Theorem 10.5 that because $(A,\Phi) = (A_0 + a,\ \Phi_0 + \phi)$ is a smooth, finite action solution to (1.4), there exists a constant $z < \infty$ such that

$$|D_A\Phi| + |F_A| < z(1+|x|)^{-2} \quad ,$$

$$0 < 1 - |\Phi| < z(1+|x|)^{-1} \quad . \qquad (9.7) \quad (5.7)$$

Hence from Theorem II.3.1, (A,Φ) has integer monopole number N' given by (1.9), and so $(A,\Phi) \in \mathscr{H}_{N'}$. By assumption $(A_0,\Phi_0) \in \mathscr{H}_N$. We prove that $N' = N$. With $b_R(x)$ defined previously, we derive by a series of integrations by parts that

$$\mathrm{Tr}\int_{\mathbb{R}^3}\left\{b_R\left(D_A\Phi\wedge F_A - D_{A_0}\Phi_0\wedge F_{A_0}\right)\right\} = -\,\mathrm{Tr}\int_{\mathbb{R}^3} db_R\wedge\left\{\phi F_A + \Phi_0 a\wedge a - D_{A_0}\Phi_0\wedge a\right\} \tag{5.8}$$

Here, Tr is the normalized trace as explained in §1. Taking absolute values of both sides in (5.8) we obtain

$$\left|\mathrm{Tr}\int_{\mathbb{R}^3} b_R\left\{D_A\Phi\wedge F_A - D_{A_0}\Phi_0\wedge F_{A_0}\right\}\right| \le$$

$$\frac{\|\nabla b\|_\infty}{R}\left\{\|\phi\|_{L_2}\|F_A\|_{L_2} + \|\Phi_0\|_\infty\|a\|^2_{L_2} + \|D_{A_0}\Phi_0\|_{L_2}\|a\|_{L_2}\right\} \tag{5.9}$$

Since $\psi = \phi + \tau_j a_j \in L_2$, the right hand side is finite, and monotonically decreasing as $R\to\infty$. The left hand side in (5.9) is bounded independently of R by $\mathscr{A}(A_0,\Phi_0) + \mathscr{A}(A,\Phi)$. The dominated convergence theorem ensures that

$$\left|\mathrm{Tr}\int_{\mathbb{R}^3}\left\{D_A\Phi\wedge F_A - D_{A_0}\Phi_0\wedge F_{A_0}\right\}\right| = 0 \tag{5.10}$$

as claimed.

IV.6. PATCHING TOGETHER MONOPOLES

We will construct explicit configurations $(A_0,\Phi_0)\in\mathscr{H}_N$ which satisfy the bounds of Theorem 2.1 when $g = su(2)$. These configurations are constructed by glueing together translations of the $N=-1$ solution of Prasad-Sommerfield, equation (1.15). The gluing is done in a singular gauge in which it is convenient to estimate $(*F_{A_0} - D_{A_0}\Phi_0)$. The fact that (A_0,Φ_0) is singular is due to the choice of gauge. This will follow from

PROPOSITION 6.1. *Let* $\{U_\alpha\}_{\alpha=1}^{\infty}$ *be a uniform open cover of* $\mathbb{R}^3$. *Let* $\hat{A}_0$ *be a connection on the vector bundle* $\mathbb{R}^3 \times g = E$ *and let* $\hat{\Phi}_0$ *be a section of* E. *Suppose there exist gauge transformations* $\{g_\alpha : U_\alpha \to G\}_{\alpha=1}^{\infty}$ *such that in each* U_α, *both* $g_\alpha \hat{A}_0 g_\alpha^{-1} + g_\alpha dg_\alpha^{-1}$ *and* $g_\alpha \hat{\Phi}_0 g_\alpha^{-1}$ *are smooth. Then there exists a gauge transformation* $g: \mathbb{R}^3 \to SU(2)$ *such that* $g\hat{A}_0 g^{-1} + g dg^{-1}$ *and* $g\hat{\Phi}_0 g^{-1}$ *are smooth.*

Proof. Set $A_\alpha = g_\alpha \hat{A}_0 g_\alpha^{-1} + g_\alpha dg_\alpha^{-1}$ and $\Phi_\alpha = g_\alpha \hat{\Phi}_0 g_\alpha^{-1}$. In each non-trivial intersection $U_\alpha \cap U_\beta$,

$$A_\beta = g_{\beta\alpha} A_\alpha g_{\beta\alpha}^{-1} + g_{\beta\alpha} dg_{\beta\alpha}^{-1}$$

$$\Phi_\beta = g_{\beta\alpha} \Phi_\alpha g_{\beta\alpha}^{-1}$$

with

$$g_{\beta\alpha} = g_\beta g_\alpha^{-1} \quad . \tag{6.1}$$

The open cover $\{U_\alpha\}_{\alpha=1}^{\infty}$ and the transition functions $\{g_{\alpha\beta} : U_\alpha \cap U_\beta \to G\}$ define a principal G bundle, P' over $\mathbb{R}^3$ (see Kobayashi and Nomizu [8], page 50). Denote by E' the vector bundle associated to P' via the adjoint representation. The set $\{\Phi_\alpha\}_{\alpha=1}^{\infty}$ is a section of E' and $\{A_\alpha\}_{\alpha=1}^{\infty}$ is connection on E' (see Chapter V.4).

LEMMA 6.2. *The bundles* P' *and* E' *are* C^∞ *fiber bundles over* $\mathbb{R}^3$.

Proof. This follows from the fact that the transition functions $g_{\beta\alpha}$ are smooth. In fact in $U_\alpha \cap U_\beta$,

$$dg_{\beta\alpha} = g_{\beta\alpha} A_\alpha - A_\beta g_{\beta\alpha} \quad . \tag{6.2}$$

Hence $|\nabla g_{\beta\alpha}| \le |A_\beta| + |A_\alpha|$ so $g_{\beta\alpha}$ has bounded derivatives in every open ball $B \subsetneq U_\alpha \cap U_\beta$. This implies that $g_{\beta\alpha}$ is continuous. Differentiating (6.2) and repeating the argument we conclude that in $U_\alpha \cap U_\beta$

$$|\nabla\nabla g_{\beta\alpha}| \le |\nabla g_{\beta\alpha}|(|A_\alpha| + |A_\beta|) + |\nabla A_\alpha| + |\nabla A_\beta| \quad . \tag{6.3}$$

Thus $\nabla g_{\beta\alpha}$ is continuous. Bootstrapping in this way, we conclude that $g_{\beta\alpha} \in C^{\infty}(U_\alpha \cap U_\beta; G)$ as claimed.

Continuing the proof of Proposition 6.1, let $P = \mathbb{R}^3 \times G$ be the canonical flat principal G bundle. It is well known that any two smooth principal G bundles over Euclidean space are C^{∞} isomorphic (See Chapter V.6, in particular Theorems V.6.1. See also Steenrod [18].) This means that there exists a smooth isomorphism $\theta: P' \to P$, $\theta: E \to E'$ which commute with the action of G and the projections (Definition V.4.1). The isomorphism θ is equivalent to a set of smooth maps $\{\theta_\alpha: U_\alpha \to G\}$ which have the property that where $U_\alpha \cap U_\beta \neq \emptyset$,

$$\theta_\alpha g_{\alpha\beta} = \theta_\beta \quad . \tag{6.4}$$

See Chapter V.4.

Define in each U_α the fields

$$\begin{aligned} \tilde{A}_\alpha &= \theta_\alpha A_\alpha \theta_\alpha^{-1} + \theta_\alpha d\theta_\alpha \quad , \\ \tilde{\Phi}_\alpha &= \theta_\alpha \Phi_\alpha \theta_\alpha^{-1} \quad . \end{aligned} \tag{6.5}$$

Then $\tilde{A}_\alpha$, $\tilde{\Phi}_\alpha$ are smooth in U_α. What is more, in U_α,

$$\begin{aligned} \tilde{A}_\alpha &= \theta_\alpha g_\alpha \hat{A}_0 (\theta_\alpha g_\alpha)^{-1} + \theta_\alpha g_\alpha d(\theta_\alpha g_\alpha)^{-1} \quad , \\ \tilde{\Phi}_\alpha &= \theta_\alpha g_\alpha \hat{\Phi}_0 \theta_\alpha g_\alpha \quad . \end{aligned} \tag{6.6}$$

But from (6.4) and (6.1), where $U_\alpha \cap U_\beta \neq \emptyset$,

$$\theta_\alpha g_\alpha = \theta_\beta g_\beta \quad . \tag{6.7}$$

Thus

$$\tilde{A}_\alpha = \tilde{A}_\beta \quad \text{and} \quad \tilde{\Phi}_\alpha = \tilde{\Phi}_\beta \, , \qquad x \in U_\alpha \cap U_\beta . \tag{6.8}$$

In summary, define a smooth connection and Higgs field by

$$(\tilde{A},\tilde{\Phi}) \equiv (\tilde{A}_\alpha,\tilde{\Phi}_\alpha) \quad , \qquad x \in U_\alpha \quad ,$$

and a gauge transformation

$$g \equiv \theta_\alpha g_\alpha \quad , \qquad x \in U_\alpha \quad . \tag{6.9}$$

It follows from (6.7-8) that this is consistent. Further $(\tilde{A},\tilde{\Phi})$ is smooth on $\mathbb{R}^3$ and

$$(\tilde{A},\tilde{\Phi}) = (g\hat{A}_0 g^{-1} + g^{-1}dg, g\hat{\Phi}_0 g^{-1})$$

as claimed.

IV.7. THE APPROXIMATE SOLUTIONS

In order to construct configurations (A_0,Φ_0) that satisfy the bounds given in Theorem 2.1, we rewrite the $N=-1$ solution (1.15) in a gauge in which Φ points in a constant direction in $\mathfrak{su}(2)$. We do this construction only for $G=SU(2)$. For other groups, we refer the reader to §18 and also [19]. Using polar coordinates (r,θ,χ) on $\mathbb{R}^3$ define

$$u = \cos\theta/2 + i\sin\theta/2(-\sin\chi\sigma^1 + \cos\chi\sigma^2) \quad , \tag{7.1}$$

where σ^1, σ^2 and σ^3 are the Pauli matrices. (See §1.)

Denoting by $(A(0),\Phi(0))$ the configuration in (1.15), we have

$$\hat{\Phi}(0) = u\Phi(0)u^{-1} = \left(\coth r - \frac{1}{r}\right) i/2\,\sigma^3 \quad ,$$

$$\begin{aligned}\hat{A}(0) = uA(0)u^{-1} + udu^{-1} &= (1-\cos\theta)d\chi\ i/2\,\sigma^3 \\ &+ \frac{r}{\sinh r}\,\frac{i}{2}\Big\{(\sin\chi\sigma^1 - \cos\chi\sigma^2)d\theta \\ &\qquad + (\cos\chi\sigma^1 + \sin\chi\sigma^2)\sin\theta\, d\chi\Big\} \quad .\end{aligned} \tag{7.2}$$

The configuration $(\hat{A}(0),\hat{\Phi}(0))$ fails to be smooth at the origin $r=0$, and along the half line

$$s_0 = \{(r,\theta,\chi)\,|\cos\theta = -1\} \quad . \tag{7.3}$$

The half line s_0 is called the "Dirac string" by physicists.

The approximate solutions we construct below depend on $3N$ parameters to specify N distinct points in $\mathbb{R}^3$ and an extra parameter $R>0$. It is an artifact of this construction that the N points and the parameter R cannot be chosen independently.

<u>Remark</u>. Suppose $R>0$ and N points $\{x_1,\ldots,x_N\}\in\mathbb{R}^3$ are given such that

$$d \equiv \min_{i\neq j} |x_i - x_j| > 8NR \tag{7.4}$$

but they are otherwise chosen arbitrarily. The condition $d>8NR$ implies two facts. First, the closed balls $\bar{B}_{4R}(x_k)$ $(k=1,\ldots,N)$ are disjoint. Second, for each $k\in\{1,\ldots,N\}$ it is possible to find a half line s_k running from x_k to ∞ and disjoint from $\cup_{i\neq k}\bar{B}_{4R}(x_i)$ (see Figure 7.1).

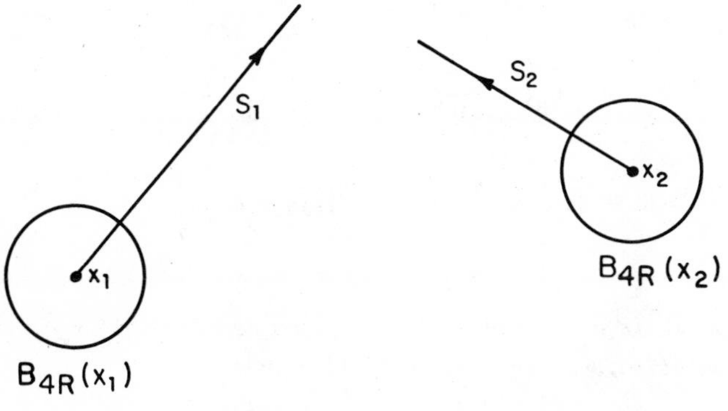

Figure 7.1. Dirac strings and balls B_{4R}.

The function $b_R(x)$ is defined in (VI.3.6). Define

$$b_R(k)(x) = b_R(x - x_k) \quad , \tag{7.5}$$

and

$$W_R(k)(x) = \prod_{\substack{i \neq k \\ i=1}}^{N} (1 - b_R(i)); \qquad k = 1,\dots,N . \tag{7.6}$$

DEFINITION 7.1. *The field configuration* $(\hat{A}_0(R,d;x_1,\dots,x_N),$ $\hat{\Phi}_0(R,d;x_1,\dots,x_N))$ *is defined for all* $R > 0$ *and* $(x_1,\dots,x_N) \in S_N(\mathbb{R}^3)$ *(the* N*-fold symmetric product of* $\mathbb{R}^3$*), subject to the constraint*

(i) $\min_{i \neq j} |x_i - x_j| \geq d > 8NR$,

by

$$\hat{\Phi}_0(R,d;x_1,\dots,x_N) = \left\{ 1 - \sum_{j=1}^{N} \left(\frac{1}{r_j} W_{2R}(j) + (1-\coth r_j)\, b_R(j) \right) \right\} \frac{i}{2} \sigma^3 ,$$

$$\hat{A}_0(R,d;x_1,\dots,x_N) = \sum_{j=1}^{N} \Big\{ (1 - \cos\theta_j) W_{2R}(j)\, d\chi_j \frac{i}{2} \sigma^3$$

$$+ \frac{i}{2} b_R(j) \frac{r_j}{\sinh r_j} \quad [(\sin\chi_j \sigma^1 - \cos\chi_j \sigma^2)\, d\theta_j$$

$$+ (\cos\chi_j \sigma^1 + \sin\chi_j \sigma^2) \sin\theta_j\, d\chi_j] \Big\} . \tag{7.7}$$

Here (r_k, θ_k, χ_k) *is a polar coordinate system centered at* x_k *with positive orientation, such that the half line*

$$s_k = \{(r_k, \theta_k, \chi_k) \mid \cos\theta_k = -1\} \tag{7.8}$$

satisfies

$$s_k \cap \bigcup_{\substack{i\neq k\\ i=1}}^{N} \bar{B}_{4R}(x_i) = \emptyset \qquad \textit{for} \qquad k=1,\dots,N . \tag{7.9}$$

In general, the dependence on the parameters $R, x_1,\dots,x_N$ will be suppressed. Notice that in each ball $B_R(x_k)$, $k=1,\dots,N$, the configuration $(\hat{A}_0,\hat{\Phi}_0)$ is, up to translation and rotation, identical to the $N=-1$ Prasad-Sommerfield solution in the singular gauge $(\hat{A}(0),\hat{\Phi}(0))$ in (7.2).

It is convenient to define open sets

$$V = \mathbb{R}^3 \setminus \bigcup_{j=1}^{N} \bar{B}_{2R}(x_j) ,$$

$$\tilde{V} = \mathbb{R}^3 \setminus \bigcup_{j=1}^{N} \bar{B}_{3.9R}(x_j) . \tag{7.10}$$

In the set V, the fields $(\hat{A}_0,\hat{\Phi}_0)$ are abelian and in the set $\tilde{V}$, equation (1.14) is satisfied exactly. The fields $(\hat{A}_0,\hat{\Phi}_0)$ are smooth except for the points $\{x_1,\dots,x_N\}$ and the half lines $\{s_1,\dots,s_N\}$. These singularities are due to the choice of gauge as we now prove.

PROPOSITION 7.2. *The configuration* $(\hat{A}_0,\hat{\Phi}_0)$ *of Definition* 7.1 *is gauge equivalent to a smooth configuration* (A_0,Φ_0) *on* $\mathbb{R}^3$.

Proof. We show that the conditions of Proposition 6.1 are satisfied. In the open set V, $\hat{\Phi}_0$ is smooth and

$$\hat{A}_0 = \sum_{j=1}^{N} (1-\cos\theta_j) W_{2R}(j)\, d\chi_j \,\frac{i}{2}\,\sigma^3 . \tag{7.11}$$

The connection is smooth except on the set of strings $\{s_j\}_{i=1}^{N}$. The gauge transformation

$$g_j = \exp(i\chi_j\sigma^3) \tag{7.12}$$

does the following. It leaves Φ_0 the same, while $g_j\hat{A}_0 g_j^{-1} + g_j dg_j^{-1}$ is smooth in V except on the set $\{s_i\}_{i\neq j} \cup \{\bar{s}_j\}$, where

$$\bar{s}_j = \{(r_j,\theta_j,x_j) \,|\, \cos\theta_j = +1\} \quad . \tag{7.13}$$

Thus the j-th string is inverted. Note that condition (7.9) is necessary for this last statement to be true. Because $g_i g_j = g_j g_i$, each string in the set $\{s_k\}_{k=1}^N$ can be inverted individually.

DEFINITION 7.3. *Let* $\hat{A}_0$ *be the connection of Definition* 7.1. *Let* $g = g_{m_1}\cdots g_{m_r}$, $(r \le N)$ *with* $m_i \in \{1,\ldots,N\}$ *and* g_{m_i} *defined by* (7.12). *With* $\hat{A}_g = g\hat{A}_0 g^{-1} + gdg^{-1}$, *define* $s_j(\hat{A}_g)$ *to be the singular half line of* $\hat{A}_g$ *with end point* x_j. *Here* $j = 1,\ldots,N$.

We summarize the above discussion in the following lemma.

LEMMA 7.4. *Let* $\hat{A}_0$ *be as in Definition* 7.1. *Let* $B_{R/10} \subset V$ *be an open ball of radius* $R/10$. *There exists a gauge transformation* $g = g_{m_1}\cdots g_{m_r}$ $(r \le N)$ *as in Definition* 7.1 *such that*

$$\mathrm{dist}(B_{R/10}, s_j(\hat{A}_g)) \ge \frac{1}{2}\,\mathrm{dist}(B_{R/10}, x_j) \quad , \quad j = 1,\ldots,N. \tag{7.14}$$

Henceforth, when dealing with the configuration $(\hat{A}_0,\hat{\Phi}_0)$ in any ball $B_{R/10} \subset V$, *we implicitly assume that a gauge transformation of the type used in Lemma* 7.4 *has been made so that* (7.14) *holds.*

In translation, Lemma 7.4 states that by a series of gauge transformations of the form (7.12) we can arrange all the Dirac strings to point away from $B_{R/10}$.

To complete the proof of Proposition 7.2 we exhibit gauge transformations which make $(\hat{A}_0,\hat{\Phi}_0)$ smooth in the open balls $B_{2.1R}(x_k)$, $k = 1,\ldots,N$.

In the ball $B_{2.1R}(x_k)$,

$$\hat{\Phi}_0 = \left[1 - \frac{1}{r_k} + (\coth r_k - 1) b_R(k)\right] \frac{i}{2}\sigma_3$$

$$\hat{A}_0 = \frac{i}{2}(1-\cos\theta_k) d\chi_k \sigma^3 + \frac{i}{2} b_R(k) \frac{r_k}{\sinh r_k} [(\sin\chi_k \sigma^1 - \cos\chi_k \sigma^2) d\theta_k$$

$$+ (\cos\chi_k\sigma^1 + \sin\chi_k\sigma^2)\sin\theta_k d\chi_k] \tag{7.15}$$

One can check that gauge transformation by

$$\hat{g}_k = \cos\frac{1}{2}\theta_k - i\sin\frac{1}{2}\theta_k(-\sin\chi_k\sigma_1 + \cos\chi_k\sigma_2) \tag{7.16}$$

renders $(\hat{A}_0, \hat{\Phi}_0)$ smooth. The proof of Proposition 7.2 is now complete.

With $(\hat{A}_0, \hat{\Phi}_0)$ as in Definition 7.1 and (A_0, Φ_0) the smooth configuration gauge equivalent to $(\hat{A}_0, \hat{\Phi}_0)$, let G_0 and $\mathcal{D}_0^{\dagger}$ be as defined in (2.8) and (2.14).

PROPOSITION 7.5. *Let* (A_0, Φ_0) *be a smooth configuration which is gauge related to* $(\hat{A}_0(R,d;x_1,\dots,x_N), \hat{\Phi}_0(R,d;x_1,\dots,x_N))$. *There exists a constant* $c_N < \infty$ *such that*

$$\|G_0\|_{L_p} + \|\mathcal{D}_0^{\dagger} G_0\|_{L_p} \le c_N R^3 \left(\frac{1}{d} + e^{-R}\right) \tag{7.17}$$

for $p = 6/5, 3/2, 2$. *The constant* c_N *is independent of* R *if* $d > 1$, *and is a function of* N *only.*

<u>Proof</u>. In the sets $\tilde{V}$ and $B_R(x_i)$, $i = 1,\dots,N$,

$$|G_0| = |\mathcal{D}_0^{\dagger} G_0| = 0 \quad . \tag{7.18}$$

It remains to estimate $|G_0|$ and $|\mathcal{D}_0^{\dagger} G_0|$ in the annuli $B_{4R}(x_i) \setminus \bar{B}_R(x_i)$, $i = 1,\dots,N$. In the open set $B_{4R}(x_i) \setminus \bar{B}_{1.9R}(x_i)$, $W_{2R}(i) = 1$ and $W_{2R}(j) \equiv W_{2R}$ for $j \neq i$. Let $B_{R/10} \subset B_{4R}(x_i) \setminus \bar{B}_{1.9R}(x_i)$ be an open ball of radius $R/10$. In $B_{R/10}$, the fields (A_0, Φ_0) are gauge equivalent (see Lemma 7.4) via a smooth gauge transformation to

$$\tilde{\Phi}_0 = \left(1 - \frac{1}{r_i} - \sum_{\substack{j=1 \\ j\neq 1}}^{N} \frac{W_{2R}}{r_j}\right) \frac{i}{2}\sigma^3$$

$$\tilde{A}_0 = (1-\cos\theta_i)d\chi_i \frac{i}{2}\sigma^3 + \sum_{\substack{j=1 \\ j\neq 1}}^{N} (1-\cos\theta_j)d\chi_j W_{2R}\frac{i}{2}\sigma^3 \quad . \tag{7.19}$$

These fields are abelian. They are exact solutions to (1.14) up to terms of order d^{-1}. In fact,

$$|G_0|(x) \leq z_1 \sum_{\substack{j=1 \\ j\neq 1}}^{N} \frac{1}{r_j} |\nabla W_{2R}| \quad ,$$

$$|\mathscr{D}_0^{\dagger} G_0|(x) \leq z_2 \sum_{\substack{j=1 \\ j\neq 1}}^{N} \left(\frac{1}{r_j}|\nabla\nabla W_{2R}| + \frac{1}{r_j^2}|\nabla W_{2R}|\right) \quad ,$$

$$\text{for} \quad x \in B_{4R}(x_i) \setminus B_{1.9R}(x_i) \ . \tag{7.20}$$

Here, $z_1, z_2 < \infty$ are independent of (A_0, Φ_0). Using the scaling properties of $b_R(x)$, (see Lemma VI.3.2) and (i) of Definition 7.1 we conclude that for $d, R > 1$ and $x \in B_{4R}(x_i) \setminus B_{1.9R}(x_i)$,

$$|G_0|(x) + |\mathscr{D}_0^{\dagger} G_0|(x) \leq z_3(N-1)\frac{1}{Rd} \quad , \tag{7.21}$$

where $z_3 < \infty$ is independent of (A_0, Φ_0).

In the sets $B_{2.1R}(x_i) \setminus \bar{B}_R(x_i)$ the configuration (A_0, Φ_0) is gauge equivalent by a smooth gauge transformation to

$$\tilde{\Phi}_0 = \left(\coth r_i - \frac{1}{r_i}\right) \frac{(x-x_i)^k}{r_i} \frac{i}{2} \sigma^k$$

$$- (1 - b_R(i))(\coth r_i - 1) \frac{(x-x_i)^k}{r_i} \frac{i}{2} \sigma^k$$

$$\tilde{A}_0 = \left(\frac{1}{\sinh r_i} - \frac{1}{r_i}\right) \varepsilon^{mjk} \frac{(x-x_i)^j}{r_i} \frac{1}{2} \sigma^k \, dx^m$$

$$- (1 - b_R(i)) \frac{1}{\sinh r_i} \varepsilon^{mjk} \frac{(x-x_i)^j}{r_i} \frac{i}{2} \sigma^k \, dx^m \tag{7.22}$$

Comparing (1.15) and (7.22) we see that in $B_{2.1R}(x_i) \setminus \bar{B}_R(x_i)$ the field configuration $(\tilde{A}_0, \tilde{\Phi}_0)$ is an exact solution up to terms of order e^{-R}. These are the terms proportional to $(1 - b_R(i))$. A calculation gives

$$|G_0|(x) \le z_4 e^{-r_i} \left(1 + |\nabla b_R(i)| + \frac{1}{r_i}\right) ,$$

$$|\mathscr{D}_0^\dagger G_0|(x) \le z_5 e^{-r_i} \left(1 + |\nabla b_R(i)| + \frac{1}{r_i} + |\nabla\nabla b_R(i)| + \frac{1}{r_i} |\nabla b_R(i)| + \frac{1}{r_i^2}\right)$$

$$\text{for } x \in B_{2.1R}(x_i) \setminus B_R(x_i). \tag{7.23}$$

Here $z_4, z_5 < \infty$ are independent of (A_0, Φ_0).

Using the scaling properties of b_R, we infer that there exists a constant $z_6 < \infty$ which is independent of (A_0, Φ_0) and $R, d > 1$ such that in $B_{2.1R}(x_i) \setminus B_R(x_i)$

$$|G_0|(x) + |\mathscr{D}_0^\dagger G_0|(x) \le z_6 e^{-R} . \tag{7.24}$$

To compute the L_p norms, we integrate (7.21) and (7.24) over $\cup_{i=1}^N (B_{4R}(x_i) \setminus B_R(x_i))$ and obtain

$$\|G_0\|_{L_p} + \|\mathscr{D}_0^\dagger G_0\|_{L_p} \le N^2 z_7 R^{3/p}\left(\frac{1}{Rd} + e^{-R}\right) ,$$

$$\le N^2 z_7 R^3\left(\frac{1}{d} + e^{-R}\right) , \tag{7.25}$$

as claimed.

PROPOSITION 7.6. *Let* (A_0,Φ_0) *be as in Proposition* 7.5. *Then* $(A_0,\Phi_0) \in \mathscr{H}_{-N}$ *(Definition* 2.3*). Further,*

(a) $\{x \in \mathbb{R}^3 : |\Phi_0(x)| = 0\} = \{x_1,\dots,x_N\}$,

(b) $|\Phi_0(x)| > \frac{1}{20}$ *if* $\min_{j\in\{x_1,\dots,x_N\}} |x - x_j| > 1.1R$,

(c) $\|\Phi_0\|_\infty = 1$.

<u>Proof</u>. Let $\rho \gg \max_{i=1,\dots,N}(|x_i|, 4R)$. Let $B_1 \subset \mathbb{R}^3 \setminus B_\rho(0)$ be a ball of radius 1. There exist a gauge in B_1 such that (A_0,Φ_0) is gauge related in B_1 to

$$\tilde{\Phi}_0 = \left(1 - \sum_{j=1}^{N} \frac{1}{r_j}\right)\frac{i}{2}\sigma^3$$

$$\tilde{A}_0 = \sum_{j=1}^{N} (1 - \cos\theta_j)dx_j \frac{i}{2}\sigma^3 . \tag{7.26}$$

In B_1, $D_{\tilde{A}_0}\tilde{\Phi}_0 = d\tilde{\Phi}_0$ and $F_{\tilde{A}_0} = d\tilde{A}_0$; explicitly,

$$D_{\tilde{A}_0}\tilde{\Phi}_0 = \sum_{j=1}^{N} \frac{1}{r_j^2} dr_j \frac{i}{2}\sigma^3 ,$$

$$F_{\tilde{A}_0} = \sum_{j=1}^{N} \sin\theta_j d\theta_j \wedge d\chi_j \frac{i}{2}\sigma^3 . \tag{7.27}$$

Their absolute values for $x \in B_1$ are

$$|F_{A_0}| = |D_{A_0}\Phi_0| = \sum_{j=1}^{N} \frac{1}{r_j^2} = \frac{N}{\rho^2} + O(\rho^{-3}) \quad .$$

Further, for $x \in B_1$,

$$(1 - |\Phi_0|) = \sum_{j=1}^{N} \frac{1}{r_j} = \frac{N}{\rho} + O(\rho^{-2}) \quad , \tag{7.28}$$

and

$$(\Phi_0, F_{A_0}) = \sum_{j=1}^{N} \sin \theta_j d\theta_j \wedge d\chi_j \left(1 - \sum_{j=1}^{N} r_j^{-1}\right) . \tag{7.29}$$

Equations (7.28-9) are gauge invariant and they are valid in any ball in $\mathbb{R}^3 \setminus B_\rho(0)$. From (7.29)

$$-\frac{1}{4\pi} \int_{|x|=\rho} (\Phi_0, F_{A_0}) = -N(1 + O(\rho^{-1})) \quad . \tag{7.30}$$

Equation (7.30) and the decay estimates of (7.28) imply that $(A_0, \Phi_0) \in \mathscr{H}_{-N}$.

To prove the second statement of the proposition break $\mathbb{R}^3$ into 4 regions and estimate $|\Phi_0|$ in each region. If $|x - x_j| < 1.1R$ then

$$|\Phi_0| = \coth r_j - \frac{1}{r_j} < 1 \qquad \text{and} \qquad |\Phi_0| = 0 \text{ iff } x = x_j \tag{7.31}$$

If $1.1R \le |x-x_j| \le 1.8R$, then

$$\frac{1}{20} < |\Phi_0| = \left| 1 - \frac{1}{|x-x_j|} + (\coth|x-x_j|-1) b_R(j) \right| < 1 \tag{7.32}$$

since in this region,

$$1 - \frac{1}{|x-x_j|} > 1 - \frac{1}{1.1R} \ge 1 - \frac{1}{1.1} > \frac{1}{20} \quad . \tag{7.33}$$

If $1.8R \le |x-x_j| < 2NR$, then by the triangle inequality and Definition 7.1, $|x-x_j| \ge |x_j-x_i| - |x-x_j| \ge 6NR$ for $i \ne j$ and $i \in \{1,2,\ldots,N\}$. Hence

$$\sum_{\substack{i\ne 1\\ i=1}}^{N} \frac{1}{|x-x_i|} W_{2R}(i) \le \frac{1}{6R} \le \frac{1}{6} \quad . \tag{7.34}$$

On the other hand where $1.8R \le |x-x_j| < 2NR$

$$|\Phi_0| = \left| 1 - \frac{1}{|x-x_j|} - \sum_{\substack{i\ne j\\ i=1}}^{N} \frac{1}{|x-x_i|} W_{2R}(i) \right| \quad . \tag{7.35}$$

Thus $5/18 < |\Phi_0| < 1$ since in this region

$$1 - \frac{1}{|x-x_j|} - \sum_{\substack{i\ne j\\ i=1}}^{N} \frac{1}{|x-x_i|} W_{2R}(i) \ge 1 - \frac{1}{1.8} - \frac{1}{6} \ge \frac{5}{18} \; . \tag{7.36}$$

Finally, if $|x-x_j| > 2NR$ for all $j \in \{1,\ldots,N\}$, then

$$\sum_{i=1}^{N} \frac{1}{|x-x_i|} W_{2R}(i) < \frac{1}{2R} \le \frac{1}{2} \quad ; \tag{7.37}$$

and this implies $\frac{1}{2} < |\Phi_0| < 1$.

Summarizing equations (7.31-7) we conclude that $\{x \in \mathbb{R}^3 : \Phi_0(x) = 0\} = \{x_1,\ldots,x_N\}$ as claimed. Further $|\Phi_0(x)| < 1$ for all $x \in \mathbb{R}^3$ and $|\Phi_0(x)| > 1/20$ for $\min_{j\in\{1,\ldots,N\}} |x-x_j| > 1.1R$. Since $\lim_{|x|\to\infty} |\Phi_0(x)| = 1$, the sup norm is

$$\|\Phi_0\|_\infty = 1.$$

This completes the proof of Proposition 7.6.

IV.8 THE EXISTENCE OF MULTIMONOPOLE SOLUTIONS

We combine the results of §7 with the theorems in §2 to prove Theorems 1.1 and 1.2.

Proof of Theorem 1.1. We are given an integer $N \geq 2$. (For $N = 1$, the solution is (1.15).) Define

$$\delta_0 = \min(1, \varepsilon_0/4) \tag{8.1}$$

where ε_0 is defined by Theorem 2.1. With c_N given by Proposition 7.5, set

$$d_0 = \max((8N)^2, (c_N 6^3 4)^2 \delta_0^{-2}) \quad . \tag{8.2}$$

For $d > d_0$ set

$$R(d) = \ln d \quad .$$

LEMMA 8.1. *Let* d_0, $R(d)$ *be as defined with* $d > d_0$. *Then for* $N \geq 2$,

(i) $d > 8NR(d)$

$$\text{(ii)} \quad c_N R(d)^3 \left(\frac{1}{d} + e^{-R(d)}\right) < \left(\frac{d_0}{d}\right)^{1/2} \frac{1}{2}\delta_0 \quad . \tag{8.3}$$

Proof. Because $d > d_0$, statement (i) is equivalent to the assertion that

$$f(d) = d - 8N \ln d > 0 \quad \text{if} \quad d > (8N)^2 . \tag{8.4}$$

Notice that for $d > 8N$, $f(d)$ is monotonically increasing since $f'(d) = 1 - 8N/d$. Further,

$$f(64N^2) = 8N(8N - 2 \ln 8N) > 0$$

for $N \geq 2$. Thus (8.4) is true. As for statement (ii), note first that

$$(\ln d)^3 < 6^3 d^{1/2} \quad , \tag{8.5}$$

since the function $f(x) = x^3 e^{-x/2}$ has its maximum on $[0,\infty]$ at $x = 6$. Hence for $x \geq 0$,

$$f(x) \leq f(6) = (6e^{-1})^3 < 6^3 \quad . \tag{8.6}$$

Therefore,

$$c_N R(d)^3 \left(\frac{1}{d} + e^{-R(d)}\right) = 2c_N \frac{(\ln d)^3}{d} < \frac{2c_N 6^3}{d^{1/2}}$$

$$= \left(\frac{d_0}{d}\right)^{1/2} \frac{2c_N 6^3}{d_0^{1/2}} \leq \left(\frac{d_0}{d}\right)^{1/2} \frac{1}{2}\,\delta_0 \quad , \tag{8.7}$$

as claimed.

PROPOSITION 8.2. *Let* $\{x_1,\ldots,x_N\}$ *satisfy* (1.19) *with* d_0 *given by* (8.2). *Let* $(A_0(x_1,\ldots,x_N),\ \Phi_0(x_1,\ldots,x_N))$ *be a smooth configuration that is gauge equivalent to the singular configuration*

$$(\hat{A}_0(R(d),d;x_1,\ldots,x_N),\ \hat{\Phi}_0(R(d),d;x_1,\ldots,x_N))$$

of Definition 7.1. *Here* $R(d) = \ln d$. *Then* (A_0,Φ_0) *satisfies the conditions of Theorem* 2.1 *and in particular* (2.14).

We remark that Proposition 7.2 ensures that a smooth (A_0,Φ_0) exists.

<u>Proof</u>. By Proposition 7.6, $(A_0,\Phi_0) \in \mathscr{H}_{-N}$ and $\|\Phi_0\| = 1$. Therefore 1.5 is satisfied by definition. Proposition 7.5 and Lemma 8.1 give the bounds

$$\|G_0\|_{L_{3/2}} \leq c_N R^3(d)\left(\frac{1}{d} + e^{-R(d)}\right) < \frac{1}{2}\,\delta_0 < \varepsilon_0/8 \quad ,$$

and

$$\|G_0\|_{L_{6/5}} + \|G_0\|_{L_2} \le 2c_N R^3(d)\left(\frac{1}{d} + e^{-R(d)}\right) < \left(\frac{d_0}{d}\right)^{1/2} \delta_0 \le \frac{\varepsilon_0}{4} \ . \quad (8.8)$$

It is clear from (8.8) that since $\|\Phi_0\|_\infty = 1$, (A_0,Φ_0) satisfies (2.14) as well.

The conclusion is that there exists $(a,\phi) \in L_2$ such that

$$(A(x_1,\ldots,x_N),\Phi(x_1,\ldots,x_N)) \equiv (A_0(x_1,\ldots,x_N) + a, \Phi_0(x_1,\ldots,x_N) + \phi) \quad (8.9)$$

is a smooth, finite action solution to (1.14). Theorem V.1.1 states that this solution is locally gauge equivalent to a real analytic configuration. Since $(A_0,\Phi_0) \in \mathscr{H}_{-N}$, the solution $(A,\Phi) \in \mathscr{H}_{-N}$ and (1.20) is true (see Theorem 2.5). The decay estimates (1.21-3) are proved as Theorem 10.5. This completes the proof of Theorem 1.1.

Proof of Theorem 1.2. We establish the theorem by proving a series of lemmas.

LEMMA 8.3. *Let* (A_0,Φ_0) *be as in Proposition* 8.2. *Let* $(A,\Phi) = (A_0 + a, \Phi_0 + \phi)$ *be the same as in* (8.9). *There is a constant* $\mu < \infty$ *which is independent of* (A_0,Φ_0) *such that*

$$\|\psi\|_\infty < \mu(d_0/d)^{1/2} \quad , \quad (8.10)$$

where $\psi = \phi + \tau_j a_j$.

Proof. By (8.8) and Theorem 2.1 (equations (2.15-6)),

$$\|\psi\|_{L_2} + \|\psi\|_H \le c\left(\frac{d_0}{d}\right)^{1/2} \quad (8.11)$$

where the constant c is defined in Theorem 2.1.

The supnorm of ψ is estimated by Theorem 2.3. This is equation (2.18). In conjunction with (8.11) we obtain

$$\|\psi\|_\infty \le \mu_1 \|\mathscr{D}_0^\dagger G_0\|_{L_2} + \mu_1 c \left(\frac{d_0}{d}\right)^{1/2} \left(1 + \mathscr{A}(A_0,\Phi_0) + c^2 \frac{d_0}{d}\right) . \tag{8.12}$$

It follows from Proposition 7.5 and Lemma 8.1 that

$$\|\mathscr{D}_0^\dagger G_0\|_{L_2} \le c_N R(d)^3 \left(\frac{1}{d} + e^{-R(d)}\right) < \frac{1}{2}\,\delta_0 \left(\frac{d_0}{d}\right)^{1/2} . \tag{8.13}$$

Also,

$$\mathscr{A}(A_0,\Phi_0) = \frac{1}{2}\|G_0\|_{L_2}^2 + 4\pi|N| \le \frac{1}{8}\,\delta_0^2 \left(\frac{d_0}{d}\right) + 4\pi|N| . \tag{8.14}$$

Since $d_0/d < 1$, equation (8.10) follows from (8.14), (8.13) and (8.12).

LEMMA 8.4. *There exists* $1/4 \ge r_0 > 0$ *such that*

$$\coth r - \frac{1}{r} > \frac{r}{6} \qquad \text{for} \qquad r \le r_0 . \tag{8.15}$$

Proof. The function $r\coth r$ is real analytic and for r small,

$$r\coth r = 1 + \frac{r^2}{3} + O(r^4) . \tag{8.16}$$

Hence,

$$\coth r - \frac{1}{r} = \frac{r}{3} + O(r^3) . \tag{8.17}$$

The existence of r_0 in the lemma follows from (8.17).

LEMMA 8.5. *Let* (A,Φ), (A_0,Φ_0) *and* $\mu < \infty$ *be as in Lemma* 8.3. *Let* $1/4 \ge r_0 > 0$ *be as in Lemma* 8.4 *and let* d_0 *be defined by* (8.2). *Suppose that* d *satisfies, in addition to* (1.19),

$$d > d_1 \equiv \left(36\mu^2/r_0^2\right) d_0 . \tag{8.18}$$

Then

$$|\Phi(x)| > 0 \quad \textit{if} \quad \min_{j\in\{1,\ldots,N\}} |x - x_j| \ge r_0 \left(\frac{d_1}{d}\right)^{1/2} \equiv R . \tag{8.19}$$

Proof. For any $x \in \mathbb{R}^3$,

$$|\Phi(x)| \geq |\Phi_0(x)| - |\phi(x)| \geq |\Phi_0(x)| - \|\psi\|_\infty ,$$

thus

$$|\Phi(x)| \geq |\Phi_0(x)| - \mu\left(\frac{d_0}{d}\right)^{1/2} . \qquad (8.20)$$

From (8.18) and Lemma 8.4

$$\frac{1}{24} \geq \frac{r_0}{6} > \mu\left(\frac{d_0}{d}\right)^{1/2} . \qquad (8.21)$$

In particular, (8.21) implies that

$$|\Phi(x)| \geq |\Phi_0(x)| - \frac{1}{24} . \qquad (8.22)$$

Hence, from (b) of Proposition 7.6,

$$|\Phi(x)| > \frac{1}{120} \quad \text{if} \quad \min_{j\in\{1,\ldots,N\}} |x-x_j| > 1.1R . \qquad (8.23)$$

For $|x-x_j| < 1.1R$,

$$|\Phi_0(x)| = \coth r - \frac{1}{r} .$$

The function $\coth r - 1/r$ is monotonically increasing since its derivative is

$$\frac{1}{r^2} - \frac{1}{\sinh^2 r} > 0 .$$

From (8.21), (8.20) and Lemma 8.4,

$$|\Phi(x)| > 0 \quad \text{if} \quad \min_j |x-x_j| \geq \frac{r_0}{6} . \qquad (8.24)$$

If, however, $r = |x-x_j| < r_0/6$, then

$$|\Phi(x)| \geq \coth r - \frac{1}{r} - \mu\left(\frac{d_0}{d}\right)^{1/2} > \frac{r}{6} - \mu\left(\frac{d_0}{d}\right)^{1/2} . \qquad (8.25)$$

We conclude that

$$|\Phi(x)| > 0 \quad \text{for} \quad \min_{j\in\{1,\dots,N\}} |x-x_j| \geq 6\mu \left(\frac{d_0}{d}\right)^{1/2} = R \quad , \tag{8.26}$$

as claimed.

Lemma 8.5 establishes Statement (a) of Theorem 1.2. Statement (b) follows directly from Statements (a) and (c): By (a), $\hat{\Phi} = \Phi/|\Phi|$ is a smooth map,

$$\hat{\Phi}: \partial B_R(x_i) \to \{\sigma \in su(2) : |\sigma| = 1\} \simeq S^2 \quad . \tag{8.27}$$

The expression in (c) computes the degree of $\hat{\Phi}$ as a map from $S^2 \to S^2$. This degree is -1; in particular the degree is nontrivial. The field Φ is smooth, thus by homotopy considerations Φ must vanish at least once in $B_R(x_i)$ (see Lloyd [11]).

It remains to prove Statement (c). Define for $t \in [0,1]$

$$\Phi_t = \Phi_0 + t\phi \quad . \tag{8.28}$$

We consider $\Phi_t : [0,1] \to C^\infty(\mathbb{R}^3; su(2))$ as a smooth homotopy between Φ_0 and $\Phi = \Phi_{t=1}$. Notice that

$$|\Phi_t| \Big|_{\partial B_R(x_i)} > 0 \quad \text{for} \quad t \in [0,1] \quad . \tag{8.29}$$

By the homotopy invariance of degree [11], Φ has the same degree as $\hat{\Phi}_0$; considering both as maps from $\partial \bar{B}_R(x_i) \to \{\sigma \in su(2) : |\sigma| = 1\}$. By construction, $\hat{\Phi}_0$ has degree -1, which proves (c).

IV. (Part ii) MONOPOLE PROPERTIES, $\lambda \geq 0$

In the second half of Chapter IV we establish properties of the solutions to the second order Higgs model equations (1.4-5). Our results apply to the first order multi-monopole equations $(\lambda = 0)$ and their solutions constructed in Part i. They also apply more generally to finite action solutions to (1.4-5), with $0 \leq \lambda$. In particular we establish decay estimates and properties of the zero set $Z(\Phi)$ for Φ. For $\lambda = 0$, we establish that Φ is an order parameter.

For simplicity, we restrict attention to the case of the gauge group $G = SU(2)$. In the final sections we summarize the known results when G is a simple Lie group.

IV.9. BASIC IDENTITIES

We start from the second order equations (1.4) for A and Φ and derive some identities which prove useful in the later sections.

To simplify notation, and to stress the similarities with the abelian Higgs model on $\mathbb{R}^2$, define f, g, w by

$$f = *F_A, \qquad g = D_A\Phi, \qquad w = \frac{1}{2}(1-|\Phi|^2). \tag{9.1}$$

Both f and g are Lie algebra-valued 1-forms. The second order equations for (A,Φ) can be written

$$*D_A f = [g,\Phi] \quad , \tag{9.2a}$$

$$*D_A*g = \nabla_A^2\Phi = -\lambda w\Phi \quad . \tag{9.2b}$$

A complementary pair of equations result from the Bianchi identities:

$$*D_A g = [f,\Phi] \tag{9.3a}$$

$$*D_A * f = 0 \quad . \tag{9.3b}$$

The first result concerns the function w:

PROPOSITION 9.1. *Let* (A,Φ) *be a smooth solution to equations* (1.4). *Then* $w = 1/2(1-|\Phi|^2)$ *satisfies*

$$dw = -(\Phi,g) \tag{9.4a}$$

$$\Delta w = -|g|^2 + \lambda(1-2w)w \quad . \tag{9.4b}$$

Proof. Statement (a) follows from the definitions. For (b), take the divergence of (a) and use (9.2b).

Second order elliptic equations for f and g can be derived from (9.2-3) also. The derivations make use of the covariant derivative identities of Chapter VI.5.

To stress the similar behavior of g and f we define an $su(2)$ valued 6-vector

$$\Psi = \begin{pmatrix} f \\ g \end{pmatrix} , \qquad \Psi_i = \begin{pmatrix} f_i \\ g_i \end{pmatrix} \quad . \tag{9.5}$$

From the preceding equations we obtain

PROPOSITION 9.2. *Let* (A,Φ) *be a smooth solution to equations* (1.4). *Let* Ψ *be defined by* (9.5). *Then*

$$-\nabla_A^2 \Psi = [[\Psi,\Phi],\Phi] - \lambda\Phi \begin{pmatrix} 0 \\ (\Phi,g) \end{pmatrix} + \Lambda(\Psi) + \lambda w \begin{pmatrix} 0 \\ g \end{pmatrix} \tag{9.6}$$

where

$$\Lambda_{ij} = -2\varepsilon^{ikj} \begin{pmatrix} f_k & g_k \\ g_k & f_k \end{pmatrix} . \quad (9.7)$$

For future reference, define $\Lambda_1(\psi) = \Lambda(\psi) + \lambda w\binom{0}{g}$.

Proof. From equations (9.2-3), we derive elliptic second order equations for f,g by operating on (9.2-3) by D_A or $D_A{*}$. From (9.2a) and (9.2b) we obtain

$$*D_A{*}D_A f = [*D_A g,\Phi] - 2{*}(g\wedge g) ,$$

$$*D_A{*}D_A f = [[f,\Phi],\Phi] - 2{*}(g\wedge g) ,$$

and

$$D_A{*}D_A{*}f = 0 . \quad (9.8)$$

Hence

$$(*D_A{*}D_A f - D_A{*}D_A{*}f) = [[f,\Phi]\Phi] - 2{*}(g\wedge g) . \quad (9.9)$$

We now use Proposition VI.5.1 to relate the operator $-\Delta_A = *D_A{*}D_A - D_A{*}D_A{*}$ to the operator $-\nabla_A^2$. The resulting equation for f is

$$-\nabla_A^2 f = [[f,\Phi],\Phi]] - 2{*}(g\wedge g + f\wedge f) . \quad (9.10)$$

An equation for g similar to the one above for f is derived in the following way. Starting with (9.3a) we obtain

$$*D_A{*}D_A g = [*D_A f,\Phi] - {*}(f\wedge g + g\wedge f) ,$$

$$*D_A{*}D_A g = [[g,\Phi],\Phi] - {*}(f\wedge g + g\wedge f) .$$

Further, from (9.2b),

$$D_A{*}D_A{*}g = -\lambda\Phi dw - \lambda wg = \lambda\Phi(\Phi,g) - \lambda wg . \quad (9.11)$$

$$*D_A*D_Ag - D_A*D_A*g = [[g,\Phi],\Phi] - \lambda\Phi(\Phi,g) + \lambda wg - *(f\wedge g + g\wedge f). \qquad (9.12)$$

Once again we use Proposition VI.5.1 to obtain

$$-\nabla_A^2 g = [[g,\Phi],\Phi] - \lambda\Phi(\Phi,g) + \lambda wg - 2*(f\wedge g + g\wedge f) \quad . \qquad (9.13)$$

The proposition follows from (9.10) and (9.13).

In Sections 12-15, we use (9.4) and (9.6) to derive decay inequalities for f, g, and w. For simplicity we introduce the following notation. If $\chi, \beta \in \oplus_{j=1}^n \mathfrak{su}(2)$ are two $\mathfrak{su}(2)$ valued n-vectors, we denote $(\chi,\beta) = \Sigma_{i=1}^n (\chi_i,\beta_i)$. If $a \in \mathfrak{su}(2)$ then $(a,\chi) = ((a,\chi_1),\ldots,(a,\chi_n))$ is a real valued n-vector.

The Higgs field Φ defines the following decomposition:

$$\chi_L = |\Phi|^{-2}\Phi(\Phi,\chi) \quad ,$$

$$\chi_T = \chi - \chi_L \quad . \qquad (9.14)$$

PROPOSITION 9.3. *Let* (A,Φ) *and* Ψ *be as in Proposition* 9.1.

(a) $|\Phi|^2(\nabla_{A_j}\Psi)_T = [\Phi,[g_j,\Psi]] - [\Phi,\nabla_{A_j}[\Phi,\Psi]]$,

(b) $(\Phi,(\nabla_{A_j}\Psi)_L) = -(g_j,\Psi) + \nabla_j(\Phi,\Psi) \quad . \qquad (9.15)$

<u>Proof</u>. In both (a) and (b), use the fact that $g_j = \nabla_{A_j}\Phi$. In (a), apply the rule

$$[a,[b,c]] = b(a,c) - c(a,b) \qquad (9.16)$$

for $a, b, c \in \mathfrak{su}(2)$.

PROPOSITION 9.4. *Let* (A,Φ) *and* Ψ *be as in Proposition* 9.2. *Then*

(a) $$-\Delta(\Phi,\Psi) = -\lambda|\Phi|^2\left(\Phi, \binom{0}{g}\right) + \lambda w\left(\Phi, \binom{0}{g}\right) + \lambda w(\Phi,\Psi) - 2(\nabla_{A_j}\Phi,\nabla_{A_j}\Psi) + (\Phi,\Lambda(\Psi)) \quad . \tag{9.17}$$

(b) $$-\frac{1}{2}\Delta|[\Phi,\Psi]|^2 = -|\nabla_A[\Phi,\Psi]|^2 - 2([\Phi,\Psi],[\nabla_{A_j}\Phi,\nabla_{A_j}\Psi]) + \lambda w|[\Phi,\Psi]|^2 - |\Phi|^2|[\Phi,\Psi]|^2 + \lambda w|[\Phi,g]|^2 + |\Phi|^2(\Psi_T,\Lambda(\Psi)) \quad . \tag{9.18}$$

Proof. For any two differentiable, $su(2)$-valued n-vectors (see Chapter VI, §5)

$$a(x),b(x) \in C'(\mathbb{R}^3;\overset{n}{\oplus} su(2))$$

$$\Delta(a,b) = (\nabla_A^2 a,b) + (a,\nabla_A^2 b) + 2(\nabla_A a,\nabla_A b) \quad . \tag{9.19}$$

Applying (9.19) we obtain

$$-\Delta(\Phi,\Psi) = -(\nabla_A^2\Phi,\Psi) - (\Phi,\nabla_A^2\Psi) - 2(\nabla_{A_j}\Phi,\nabla_{A_j}\Psi) \quad . \tag{9.20}$$

Now use (9.2b) and (9.6) to infer Statement (a). As for Statement (b), we use equation (9.19) to obtain

$$-\frac{1}{2}\Delta|[\Phi,\Psi]|^2 = -|\nabla_A[\Phi,\Psi]|^2 - ([\Phi,\Psi],\nabla_A^2[\Phi,\Psi]) \quad . \tag{9.21}$$

The last term on the right hand side of (9.21) is written

$$\nabla_A^2[\Phi,\Psi] = 2[\nabla_{A_j}\Phi,\nabla_{A_j}\Psi] + [\nabla_A^2\Phi,\Psi] + [\Phi,\nabla_A^2\Psi] \quad . \tag{9.22}$$

Now use (9.2b), (9.6) and (9.16).

IV.10. DECAY PROPERTIES

Through this section, take $G = SU(2)$. The interaction between the gauge field and the Higgs field manifests itself in the large $|x|$ decay of the field strengths. The heuristic picture is discussed in Chapter I. This heuristic discussion is summarized in the theorems below. We consider first $\lambda > 0$.

THEOREM 10.1. *Let* (A,Φ) *be a smooth, finite action solution to equations* (1.4) *with* $\lambda > 0$. *Then* $|\Phi| < 1$ *and* $\lim_{|x|\to\infty} |\Phi| \to 1$ *uniformly with* $|x|$.

THEOREM 10.2. *Let* (A,Φ) *be a smooth, finite action solution to equations* (1.4) *with* $\lambda > 0$. *Given* $\varepsilon > 0$, *there exists a constant* $M(\varepsilon) < \infty$ *such that*

$$\text{(a)} \quad |(F_A)_T| + |(D_A\Phi)_T| \leq M(\varepsilon)e^{-(1-\varepsilon)|x|}.$$

$$\text{(b)} \quad |(D_A\Phi)_L| + |w| \leq M(\varepsilon)e^{-(1-\varepsilon)m|x|}, \tag{10.1}$$

where $m = m_L = \min(\lambda^{\frac{1}{2}},2)$. There exists a constant $M < \infty$ such that

$$\text{(c)} \quad |(F_A)_T| \leq M(1+|x|^2)^{-1} \quad . \tag{10.2}$$

When $\lambda = 0$, similar results to those obtained above are found. The following theorem justifies the boundary condition $|\Phi| \to 1$, as explained in §1.

THEOREM 10.3. *Let* (A,Φ) *be a smooth solution to equations* (1.4) *with* $\lambda = 0$. *Then*

$$\lim_{|x|\to\infty} |\Phi(x)| \to c, \quad \textit{uniformly with } |x|, \tag{10.3}$$

where c *is a constant. Further for all* $x \in \mathbb{R}^3$, *either* $|\Phi| \equiv c$, *or else*

$$|\Phi(x)| < c \quad . \tag{10.4}$$

In Section 1, we saw

COROLLARY 10.4. *If the constant* c *in Theorem* 10.3 *is zero, then the solution* (A,Φ) *is gauge equivalent to zero.*

THEOREM 10.5. *Let* (A,Φ) *be a smooth, finite action solution to equations* (1.4) *with* $\lambda = 0$, *and suppose* $|\Phi| \to 1$ *uniformly as* $|x| \to \infty$. *Given* $\varepsilon > 0$, *there exists a constant* $M(\varepsilon) < \infty$ *such that*

$$\text{(a)} \qquad |(F_A)_T|, \quad |(D_A\Phi)_T| \leq M(\varepsilon)e^{-(1-\varepsilon)|x|} . \tag{10.5}$$

Further there exists a constant $z < \infty$ *such that*

$$\text{(b)} \qquad |(F_A)_L|, \quad |(D_A\Phi)_L| \leq z(1+|x|^2)^{-1}. \tag{10.6}$$

$$\text{(c)} \qquad 1-|\Phi| \leq z(1+|x|^2)^{-1/2} . \tag{10.7}$$

A bound on $\|F_A\|_\infty$ and $\|D_A\Phi\|_\infty$ is obtained which depends solely on the action $\mathscr{A}(A,\Phi)$. This is a consequence of Proposition V.7.2 of the next chapter and is derivable for $\lambda \geq 0$:

PROPOSITION 10.6. *Let* (A,Φ) *be a smooth solution to equations* (1.4) *with* $\lambda > 0$ *and* $|\Phi| \sim 1$ *as* $|x| \to \infty$. *There exists a constant* $M(\mathscr{A}(A,\Phi))$ *such that*

$$\sup_{x \in \mathbb{R}^3} (|F_A| + |D_A\Phi|) \leq M(\mathscr{A}(A,\Phi)) . \tag{10.8}$$

We state without presenting a proof here that decay bounds for fields such as $(D_A^{(I)}F_A)_{T,L}$ and $(D_A^{(I)}D_A\Phi)_{T,L}$ exists. Here $(I) = (i_1,\ldots,i_n)$ is a multi-index and $D_A^{(I)} = D_{Ai_1}\cdots D_{Ai_n}$. These bounds are obtained by differentiating $|I| (= \Sigma_{j=1}^n i_j)$ times the bounds above for $(F_A)_{T,L}$ and $(D_A\Phi)_{T,L}$, respectively; see [20].

The proofs of Theorems 10.1-10.6 are contained in the next three sections. It is convenient to prove the results in a different order than we stated them. In §11, we prove Proposition 10.6, Theorems 10.1 and 10.3.

In §§12, 13 we prove the exponential decay estimates in Theorems 10.2 and 10.5, leaving until §§14, 15 the proof of the power law decay.

IV.11. THE MAXIMUM FIELD STRENGTH

We prove that the absolute value of the Higgs field approaches its asymptotic value uniformly as $|x| \to \infty$, and furthermore $|\Phi|$ is less than this asymptotic value. In the process, we establish a pointwise bound depending on the action. Then for a smooth finite action solution to (1.4) we prove $\|\Phi\|_{L_\infty} < \infty$ and

$$\|F_A\|_{L_\infty} + \|D_A\Phi\|_{L_\infty} \leq \text{const}\ (\mathscr{A}^{1/2} + \mathscr{A}^{5/2})(1 + \|\Phi\|_{L_\infty}) \tag{11.1}$$

with the constant independent of (A,Φ). Furthermore, if $0 < \lambda$, then $\|\Phi\|_{L_\infty} = 1$.

PROPOSITION 11.1. *Let* (A,Φ) *be a smooth, finite action solution to equations* (1.4) *with* $\lambda \geq 0$. *Then* $|f|$, $|g|$ *satisfy* (11.1) *and have uniform decay as* $|x| \to \infty$.

<u>Proof</u>. We also assume here that $\|\Phi\|_{L_\infty} \leq 1$. After proving this case, we eliminate the extra assumption. It is convenient to treat f and g as a single $\mathcal{g}$-valued 6-vector Ψ, defined in (9.5). The vector Ψ satisfies an equation of the form

$$-\nabla_A^2 \Psi = \Lambda(\Psi) + v \quad . \tag{11.2}$$

Here Λ is given in (9.7) and

$$|\Lambda| \leq 4(|f|^2 + |g|^2) \quad . \tag{11.3}$$

The vector v satisfies (see (9.6))

$$|v| \leq \left(1 + \|\Phi\|^2_{L_\infty}\right)(1 + 2\lambda)|\Psi| \in L_2 \tag{11.4}$$

Theorem V.7.1 is applicable under these circumstances. As a result of the theorem, $|f|$, $|g| \in L^1_6$; hence they have uniform decay as $|x| \to \infty$ (by Proposition III.7.5). Further, a direct result of Proposition V.7.2 is Proposition 10.6 and (11.1).

We next eliminate our assumption on $\|\Phi\|_{L_\infty}$. First suppose $0 < \lambda$. Then $w \in L_2$ and satisfies, see (9.4),

$$\Delta w \leq \lambda(1 - 2w)w \quad . \tag{11.5}$$

From the maximum principle, Chapter VI.3, we infer either $w \equiv 0$ or $w > 0$, and $\|\Phi\|_{L_\infty} \leq 1$.

We next prove for $\lambda = 0$ that $|\Phi| \leq c < \infty$, where c is constant. If $c \leq 1$, the previous proof holds. If $c > 1$, we perform the scale transform (1.8) to produce another solution (A', Φ') with $c' = 1$ and $\mathscr{A}' = c^{-1}\mathscr{A}$. Then (11.1) holds.

The strong maximum principle (Proposition VI.3.1) states that w can have no non-positive minimum. This does not rule out $|\Phi|$ becoming large and $w \to -\infty$ as $|x| \to \infty$. This possibility is eliminated by investigating the behavior of w implied by the equation

$$\Delta w = -|g|^2 \quad . \tag{11.6}$$

If $\Psi_1 = (a_{1j}, b_{1j})^3_{j=1}$ and $\Psi_2 = (a_{2j}, b_{2j})$ are two $su(2)$ valued 6-vectors, define an inner product

$$(\Psi_1, \Psi_2) = (a_{1j}, a_{2j}) + (b_{1j}, b_{2j}) \quad . \tag{11.7}$$

From (9.6) we obtain an inequality for $\lambda \geq 0$ and $\Psi = (f_j, g_j)^3_{j=1}$:

$$-(\Psi,\nabla_A^2\Psi) \leq (\Psi,\Lambda_1(\Psi)) \qquad , \tag{11.8}$$

with Λ_1 given by (9.7).

PROPOSITION 11.2. *Let* (A,Φ) *be a smooth solution to equations* (1.4) *with* $\lambda \geq 0$. *Suppose that the action* $\mathcal{A}(A,\Phi) < \infty$. *Then*

$$\| \nabla|\Psi|\|_{L_2}^2 \leq \| \nabla_A \Psi\|_{L_2}^2 \leq c(1+\mathcal{A}^2)\mathcal{A} \qquad , \tag{11.9}$$

with $c < \infty$ *independent of* (A,Φ).

Proof. The proof is an application of Proposition V.7.2 to equation (11.8). We remark that $|\Lambda_1| \in L_2$ (by (11.3)) and

$$\|\Lambda_1\|_{L_2}^2 \leq \text{const } \mathcal{A}(A,\Phi) \qquad . \tag{11.10}$$

A result of Proposition 11.2 is that $|\Psi| \in L_p$, $2 \leq p \leq 6$. This is a Sobolev inequality (cf. VI, §1 and §2).

PROPOSITION 11.3. *Let* (A,Φ) *be a smooth, finite action solution to equations* (1.4) *for* $\lambda = 0$. *Then* $|\Phi|$ *is bounded on* $\mathbb{R}^3$.

Proof. We prove equivalently that w is bounded from below. Define a function $\ell(x)$ by

$$\ell(x) = \frac{1}{4\pi}\int_{\mathbb{R}^3} dy \, \frac{|g(y)|^2}{|x-y|} \qquad . \tag{11.11}$$

An upper bound for $\ell(x)$ is given by dividing the integral (11.11) into $|x-y| \leq 1$ and $|x-y| > 1$. Then

$$0 \leq \ell(x) \leq \text{const}\left(\|g\|_{L_2}^2 + \|g\|_{L_4}^2\right) \qquad . \tag{11.12}$$

Further, $\ell(x)$ is smooth and decays uniformly to zero as $|x| \to \infty$, by Proposition VI.4.1.

The kernel $(4\pi|x-y|)^{-1}$ is the Green's function for $-\Delta$. Therefore

$$w - \ell = -h \quad , \tag{11.13}$$

and $-h$ is a smooth harmonic function on $\mathbb{R}^3$ which is bounded from below.

LEMMA 11.4. *The function* $h(x)$ *defined by* (11.3) *is a constant.*

Proof. Let $M(R)$ be the maximum value of $h(x)$ in the ball of radius R. Since h is harmonic, $M(R)$ is achieved at some x_0 satisfying $\|x_0\| = R$; this by the maximum principle (§VI.3). At x_0,

$$|\Phi|^2(x_0) = 2(M(R) - \ell(x_0)) + 1 \quad . \tag{11.14}$$

We remark that for $|x| \leq R$, sufficiently large,

$$|\nabla|\Phi|^2| \leq 2(M(R))^{1/2}|g| \quad . \tag{11.15}$$

The function $|\Phi|^2$ is assumed to be C^∞ and hence it is Hölder continuous in the ball of radius $2R$. The Hölder exponent is given by Corollary VI.2.7, equation (VI,2.27).

$$\begin{aligned} ||\Phi|^2(x) - |\Phi|^2(0)| &\leq c|x|^{1/2} \|\nabla|\Phi|^2\|_{L_6} \\ &\leq c|x|^{1/2}\, 2M(R)^{1/2} \|g\|_{L_6} \quad . \end{aligned} \tag{11.16}$$

Here c is a constant independent of (A,Φ) and R. The quantity $|\Phi|^2(0)$ is some constant. Set $x = x_0$ in (11.16). Then

$$M(R) \leq \ell(x_0) + |\Phi|^2(x_0) + 2c_1 R^{1/2} M(R)^{1/2} \|g\|_{L_6} \quad . \tag{11.17}$$

Using the triangle inequality and (11.12)

$$M(R) \leq c_4(\|g\|^2_{L_2} + \|g\|^2_{L_4}) + 2|\Phi|^2(x_0) + 4c_1^2 \, R\|g\|^2_{L_6} \quad ,$$

$$\leq c_2 + c_3 R \quad . \tag{11.18}$$

Recall that $M(R) = \sup_{|x|\leq R} h(x)$ and $h(x)$ is harmonic. Therefore h is a linear function of the coordinate x,

$$h(x) \;=\; c + b_j x^j \quad , \tag{11.19}$$

with c, b_j constants. But $h(x)$ is bounded from below, so $b_j \equiv 0$, $j = 1,2,3$ which proves the lemma.

By scaling (A,Φ) we arrange so that

$$w(x) \;=\; \frac{1}{4\pi} \int_{\mathbb{R}^3} dy \; \frac{|g|^2(y)}{|x-y|} \quad . \tag{11.20}$$

Then $|\Phi|$ is bounded on $\mathbb{R}^3$ by 1 and w decays uniformly to zero as $|x| \to \infty$. We have established that $|\Phi| < 1$.

Proof of Corollary 10.4. Since w achieves its minimum at infinity and we conclude that $w \equiv 1$. Thus $|\Phi| = |g| = 0$. But $\|g\|_{L_2} = \|f\|_{L_2}$ from Corollary II.2.2.

IV.12. TRANSVERSE FIELDS: $e^{-|x|}$ DECAY

The proof of the exponential decay estimates of Theorems 10.2 and 10.5 are straightforward, although the algebraic manipulations are cumbersome. A simpler but technically similar calculation is given in §8 and §9 of Chapter III where we derived exponential decay bounds for the field strengths in the abelian Higgs model on $\mathbb{R}^2$.

Since w decays uniformly to zero (Section 11), there exists a radius R outside of which $|\Phi| > 1/2$. For $|x| > R$ we write equation (9.6) as

$$-\nabla_A^2\Psi = -\Psi_T - \lambda\begin{pmatrix}0\\ g_L\end{pmatrix} + \Lambda(\Psi)_T + \Lambda(\Psi)_L + 2w\Psi_T + \lambda w\begin{pmatrix}0\\ g_T\end{pmatrix} + 3\lambda w\begin{pmatrix}0\\ g_L\end{pmatrix} . \tag{12.1}$$

The subscripts T, L are defined in equations (9.14). The two projections $\Psi \to \Psi_T, \Psi_L$ do not commute with ∇_A^2 or Λ and this is the technical problem that we must solve.

PROPOSITION 12.1. *Let* $\Psi = (f_i, g_i)$ *and* Λ *be given by* (9.7). *Then*

$$|(\Psi_T, \Lambda(\Psi))| \leq 8|\Psi_T|^2|\Psi_L| , \tag{12.2}$$

and

$$|\Lambda(\Psi)_L| \leq 4|\Psi_T|^2 . \tag{12.3}$$

Proof. The proposition follows from the fact that $\Lambda(\Psi)$ is a commutator and two longitudinal components commute. Explicitly,

$$\Lambda(\Psi) = -\varepsilon^{ijk}\begin{pmatrix}[f_j, f_k] + [g_j, g_k]\\ 2[f_j, g_k]\end{pmatrix} . \tag{12.4}$$

Thus

$$(\Lambda_T(\Psi_T))_T = 0 ,$$

$$(\Lambda_T(\Psi_L))_L = (\Lambda_L(\Psi_T))_L = 0 ,$$

$$(\Lambda_L(\Psi_L)) = 0 ,$$

where

$$(\Lambda_{T,L})_{ik} = -2\varepsilon^{ijk}\begin{pmatrix}f_{T,Lj} & g_{T,Lj}\\ g_{T,Lj} & f_{T,Lj}\end{pmatrix} . \tag{12.5}$$

Hence,

$$|(\Psi_T,\Lambda(\Psi))| \leq |\Psi_T||\Lambda_T||\Psi_L| + |\Psi_T||\Lambda_L||\Psi_T| \quad ,$$

$$\leq 8|\Psi_T|^2|\Psi_L| \quad . \tag{12.6}$$

This establishes (12.2). Equation (12.3) follows from the fact

$$(\Lambda(\Psi))_L = \Lambda_T(\Psi_T) \quad , \tag{12.7}$$

and the bound

$$|\Lambda_{T,L}| \leq 4|\Psi_{T,L}| \quad . \tag{12.8}$$

The more complicated technical problem occurs with the term $-\nabla_A^2\Psi$. There are two equivalent approaches. The first is to mimic the proof of exponential decay for the field strength in the abelian Higgs model (Chapter III, §9) by decomposing the Laplacian in a convenient gauge. The second approach, which we will use, is to derive differential inequalities for $|[\Phi,\Psi]|$ and $|(\Phi,\Psi)|$ directly.

The function $|[\Phi,\Psi]|$ satisfies equation (9.18) (see Proposition 9.4). For notational convenience set $\eta = [\Phi,\Psi]$.

LEMMA 12.2. *Let* (A,Φ) *and* Ψ *be as in Proposition* 9.4. *Then*

$$|(\eta,[\nabla_{A_j}\Phi,\nabla_{A_j}\Psi])| \leq 2|\Phi|^{-2}|\eta|^2|\Psi_L|^2$$
$$+ |\Phi|^{-1}|\eta||\nabla_A\eta||\Psi_L|$$
$$+ |\Phi|^{-1}|\eta|^2|(\nabla_A\Psi)_L| \quad . \tag{12.9}$$

<u>Proof</u>. Since $\eta = \eta_T$,

$$(\eta,[\nabla_{A_j}\Phi,\nabla_{A_j}\Psi]) = (\eta,[(\nabla_{A_j}\Phi)_T,(\nabla_{A_j}\Psi)_L]) ,$$
$$+ (\eta,[(\nabla_{A_j}\Phi)_L,(\nabla_{A_j}\Psi)_T]) . \quad (12.10)$$

The first term on the right hand side of (12.10) is bounded by

$$|\eta||\Psi_T||(\nabla_A\Psi)_L| = |\Phi|^{-1}|\eta|^2|(\nabla_A\Psi)_L| . \quad (12.11)$$

As for the second term, we use Statement (a) of Proposition 9.3:

$$|(\eta,[(g_j)_L,(\nabla_{A_j}\Psi)_T])| \leq |\Phi|^{-2}|(\eta,[(g_j)_L,[\Phi,[g_j,\Psi]]])|+|\Phi|^{-2}|(\eta,[(g_j)_L,$$
$$[\Phi,\nabla_{A_j}\eta]])$$
$$\leq 2|\Phi|^{-2}|\eta|^2|\Psi_L|^2 + |\Phi|^{-1}|\eta||\nabla_A\eta||\Psi_L| . \quad (12.12)$$

Using Propositions 12.1 and 12.2 and equation (9.18) we derive the following inequality:

$$-\frac{1}{2}\Delta|\eta|^2 + |\nabla_A\eta|^2 \leq -|\eta|^2+2(1+\lambda)w|\eta|^2 + 8|\eta|^2|\Psi_L|$$
$$+ 4|\Phi|^{-2}|\eta|^2|\Psi_L|^2 + 2|\Phi|^{-1}|\eta|^2|(\nabla_A\Psi)_L|$$
$$+ 2|\Phi|^{-1}|\eta||\nabla_A\eta||\Psi_L| . \quad (12.13)$$

LEMMA 12.3. *For any* $0 < \mu \leq 1$ *and all* $x > R$,

$$-\frac{1}{2}\Delta|\eta|^2 + (1-\mu)|\nabla_A\eta|^2 \leq -|\eta|^2(1-v(\mu)) \quad (12.14)$$

where

$$v(\mu) = 2(1+\lambda)w + 8|\Psi_L| + 16|\Psi_L|^2 + 16\mu^{-1}|\Psi_L|^2 + 4|(\nabla_A\Psi)_L| . \quad (12.15)$$

Proof. We use the fact that in the region of interest, $|\Phi| > 1/2$. The μ dependence comes from completing the square in the last term on the right hand side of (12.13).

PROPOSITION 12.4. *Let* (A,Φ), Ψ *be as in Proposition* 9.1. *Then* $|\nabla_A \Psi|$ *decays uniformly to zero.*

Proof. The proposition is proved by demonstrating that $|\nabla_A \Psi| \in L_6^1$ and appealing to Proposition III.7.5. Proposition 11.2 established that $|\nabla_A \Psi| \in L_2$. Differentiating both sides of (9.6) by ∇_{A_i} and commuting covariant derivatives (see Proposition VI.5.2)

$$-\nabla_A^2 \nabla_{A_i} \Psi = [[\nabla_{A_i}\Psi,\Phi],\Phi] + [[\Psi,g_i],\Phi] + [[\Psi,\Phi],g_i]$$

$$- \lambda g_i \begin{pmatrix} 0 \\ (\Phi,g) \end{pmatrix} - \lambda\Phi \begin{pmatrix} 0 \\ (g_i,g) \end{pmatrix} - \lambda\Phi \begin{pmatrix} 0 \\ (\Phi,\nabla_{A_i} g) \end{pmatrix}$$

$$+ \lambda(\Phi,g_i) \begin{pmatrix} 0 \\ g \end{pmatrix} + \lambda w \begin{pmatrix} 0 \\ \nabla_{A_i} g \end{pmatrix}$$

$$+ 2\varepsilon^{ik\ell}[f_\ell,\nabla_{A_k}\Psi] + [\varepsilon^{ik\ell}\nabla_k f_\ell,\Psi] + (\nabla_{A_i}\Lambda)(\Psi) + \Lambda(\nabla_{A_i}\Psi), \quad (12.16a)$$

where

$$(\nabla_{A_i}\Lambda(\Psi) + \Lambda(\nabla_{A_i}\Psi))_\ell = -2\varepsilon^{\ell jk} \begin{pmatrix} [\nabla_{A_i} f_j, f_k] + [\nabla_{A_i} g_j, g_k] \\ [\nabla_{A_i} f_j, g_k] + [f_j, \nabla_{A_i} g_k] \end{pmatrix}. \quad (12.16b)$$

Equations (12.16) for $\nabla_A \Psi$ satisfy the conditions of Theorem V.7.1. The conclusion of that theorem gives $|\nabla_A \Psi| \in L_6^1$ as claimed.

A corollary of Proposition 12.4 is that $v(\mu)$ defined in (12.15) has uniform decay as $|x| \to \infty$.

PROPOSITION 12.5. *Given* $\varepsilon > 0$, *there exists* $M(\varepsilon) < \infty$ *such that*

$$|\eta| \leq M(\varepsilon) e^{-(1-\varepsilon)|x|} \quad . \tag{12.17}$$

We remark that Proposition (12.5) establishes Statements (a) of Theorems 10.2 and 10.5.

Proof. In equation (12.14), choose $\mu = \varepsilon/2$. There exists $R(\varepsilon) > R$ such that

$$\sup_{|x|>R_\varepsilon} |v(\varepsilon/2)(x)| < \varepsilon/2 \quad . \tag{12.18}$$

Choose $M(\varepsilon)$ to satisfy

$$M(\varepsilon) > e^{(1-\varepsilon)R(\varepsilon)} \sup_{|x|\leq R(\varepsilon)} |\eta| \; . \tag{12.19}$$

Let $h_- \equiv \min(h,0)$ and

$$V = \operatorname{supp}\left((M(\varepsilon)e^{-(1-\varepsilon)|x|} - |\eta|)_-\right) \quad .$$

Since $\exp(-(1-\varepsilon)|x|) > 0$, the function $|\eta| > 0$ in V. By taking a slightly larger open set V' we assure that

(a) $0 < |\eta| \in C^\infty(\bar{V}')$

(b) $\partial\bar{V}'$ is smooth

(c) $$(M(\varepsilon)\cdot e^{-(1-\varepsilon)|x|} - |\eta|)\Big|_{\partial\bar{V}'} \geq 0 \quad . \tag{12.20}$$

A calculation gives for $x \in V'$,

$$\Delta|\eta|^{1+\varepsilon/2} = \frac{1}{2}(1+\varepsilon/2)|\eta|^{\varepsilon/2-1}\Delta|\eta|^2 - (1-\varepsilon^2/4)|\eta|^{\varepsilon/2-1}|\nabla|\eta||^2 \; . \tag{12.21}$$

Using (12.14) to compute $\Delta|\eta|^{1+\varepsilon/2}$ we obtain

$$-\Delta|\eta|^{1+\varepsilon/2} \leq -(1+\varepsilon/2)(1-\varepsilon/2)|\eta|^{1+\varepsilon/2} \quad , \quad x \in V' \quad . \tag{12.22}$$

Together equations (12.20), (12.22) and Proposition III.7.2 imply that for $|x| \in V'$

$$|\eta|^{1+\varepsilon/2} \leq M(\varepsilon)^{1+\varepsilon/2} \exp(-(1+\varepsilon/2)(1-\varepsilon)|x|) \quad . \tag{12.23}$$

But (12.23) is true by definition for $x \notin V'$ which establishes the proposition.

IV.13. $\lambda > 0$; DECAY OF $1-|\Phi|$ AND $|g_L|$

The purpose of this section is to derive the bounds given in Statement (b) of Theorem 10.2. The derivation takes three steps. Step one derives a bound on $|g_L|$ by $\exp(-m_1|x|)$ where $m_1 = \min(\lambda,1)$. Step two uses this result to bound w by $\exp(-m|x|)$, with m given by Theorem 10.2. We then differentiate the bound on w to bound $|g_L|$ by $\exp(-m|x|)$ also.

PROPOSITION 13.1. *Let* (A,Φ) *be a smooth, finite action solution to* (1.4) *with* $\lambda > 0$. *Given* $0 < \varepsilon < 1$, *there exists* $M(\varepsilon)$ *such that*

$$|g_L| \leq M(\varepsilon)e^{-(1-\varepsilon)m_1|x|} \tag{13.1}$$

where

$$m_1 = \min(\lambda^{1/2},1) \quad .$$

<u>Proof</u>. We use equation (9.17). Let $u = (\Phi,g)$. Then

$$-\Delta u = -\lambda u + 4\lambda wu + (\Phi, P_g\Lambda) - 2(g_j, \nabla_{A_j} g)$$

where

$$P_g = \begin{pmatrix} 0 & 0 \\ 0 & I \end{pmatrix} . \tag{13.2}$$

We will apply Corollary III.7.6. As in §12 we specify $R < \infty$ outside of which $|\Phi| > 1/2$.

For the last term in (13.2), we have for $|x| > R$,

$$\begin{aligned}(g_j, \nabla_{A_j} g) &= |\Phi|^{-2}(\Phi, g_j)(\Phi, \nabla_{A_j} g) + ((g_j)_T, (\nabla_{A_j} g)_T), \\ &= |\Phi|^{-2}(\Phi, g_j)(\nabla_j u - (g_j, g)) + ((g_j)_T, (\nabla_{A_j} g)_T), \\ &= |\Phi|^{-2}(\Phi, g_j)\nabla_j u - |\Phi|^{-2}|g_L|^2 u - |\Phi|^{-2}(\Phi, g_j)((g_j)_T, (g)_T) \\ &\quad + ((g_j)_T, (\nabla_{A_j} g)_T) . \end{aligned} \tag{13.3}$$

Here, the second line follows from the first by using Proposition 9.3. Each component u_i satisfies

$$-\Delta u_i + 2b^k \nabla_k u_i + \lambda u_i(1-v) = q_i , \qquad |x| > R , \tag{13.4}$$

where

$$b^k = |\Phi|^{-2}(\Phi, g_k)$$

$$v = 2(2w + \lambda^{-1}|\Phi|^{-2}|g_L|^2)$$

$$q_i = (\Phi, P\Lambda_i) + 2|\Phi|^{-2}(\Phi, g_j)((g_j)_T, (g_i)_T) - 2((g_j)_T(\nabla_{A_j}, g_i)_T) . \tag{13.5}$$

LEMMA 13.2. *Given* $\varepsilon > 0$, *there exists* $R(\varepsilon) < \infty$ *and* $M_1(\varepsilon) < \infty$ *such that the following is true:*

(a) $\sup_{|x|>R_\varepsilon} |b| < \varepsilon^2/4$,

(b) $\sup_{|x|>R_\varepsilon} v < \varepsilon/4$,

(c) $|q| \leq M_1(\varepsilon)e^{-(1-\varepsilon/4)|x|}$. (13.6)

Proof. Statements (a) and (b) follow from the uniform decay of w and $|g|$ (see Section 11). As for statement (c), use Proposition 12.1 for $|\Lambda_L|$, then Proposition 12.5 and the fact that $|\nabla_A g|$ is bounded (see Proposition 12.4).

To prove Proposition 13.1, apply Proposition III.7.4 to (13.5) for $|x| > R(\varepsilon)$. The corollary asserts that there exists $M_0(\varepsilon) < \infty$ such that

$$|u_i| \leq M_0(\varepsilon/4)e^{-\bar{m}(1-\varepsilon/4)|x|}$$

with

$$\bar{m} = \min((\lambda + \varepsilon^4/16)^{1/2} - \varepsilon^2/4, (1 - \varepsilon/4)) . \tag{13.7}$$

The proposition follows upon noting that if $\lambda^{1/2} > \varepsilon$, then

$$(\lambda + \varepsilon^4/16)^{1/2} - \varepsilon^2/4 \geq \lambda^{1/2}(1 - \varepsilon/\lambda^{1/2}\varepsilon/4) \geq \lambda^{1/2}(1 - \varepsilon/4) . \tag{13.8}$$

Hence with $M(\varepsilon) = M_0(\varepsilon/4)$,

$$|u_i| \leq M(\varepsilon)e^{-(1-\varepsilon/4)(1-\varepsilon/4)m_1|x|} \leq M(\varepsilon)e^{-(1-\varepsilon)m_1|x|} \tag{13.9}$$

as claimed.

PROPOSITION 13.3. *Let* (A,Φ) *be a smooth, finite action solution to equations* (1.4) *with* $\lambda > 0$. *Then* w *satisfies the bound given in* (10.1b).

Proof. By Proposition 13.1 and Proposition 12.5, for any $\delta > 0$, there exists $M(\delta) < \infty$ such that

$$|g|^2 \leq M(\delta)e^{-(1-\varepsilon)\mu|x|} \quad , \tag{13.10}$$

with $\mu = \min(2\lambda^{\frac{1}{2}}, 2)$. Now use equation (9.4) and Proposition III.7.4.

Next we bound $|g_L| = |\Phi|^{-1}dw$. Let $G_\lambda(x-y) = (4\pi|x-y|)^{-1}\exp(-\lambda^{\frac{1}{2}}|x-y|)$ be the Green's function for $(-\Delta + \lambda)$, $\lambda > 0$. By (9.4),

$$w(x) = \int G_\lambda(x-y)(|g(y)|^2 + 2\lambda w(y)^2)\, dy \ .$$

Differentiating,

$$(\Phi,g)(x) = -\int \nabla_x G_\lambda(x-y)(|g(y)|^2 + 2\lambda w(y)^2)\, dy \ .$$

Standard convolution estimates yield the decay (10.16) for $|g_L|$, cf. Chapter III.9.

IV.14 AN ITERATIVE METHOD FOR $|x|^{-p}$ DECAY

In nonlinear problems, the conditions specified in Chapter VI, §4 which are sufficient to establish a power law decay for solutions to Laplace's equation may not immediately apply. In this section we give an iterative technique which gives power law decay for certain nonlinear problems. The basic idea is that when the solution u has a power law decay $|x|^{-p}$, the powers $|u|^\alpha$, $\alpha > 1$, have faster decay $|x|^{-p\alpha}$. For certain equations, we can iterate this improvement to yield $|x|^{-p\alpha n}$ decay, n integer, up to some limiting power law $|x|^{-q}$. In the first proposition, for example, we start with $|x|^{-1}$ decay and an L_p bound, and we iterate up to $|x|^{-2}$ decay.

Consider a vector-valued solution $u = (u_1, u_2, \ldots, u_n)$ to the equation on $\mathbb{R}^3$,

$$\Delta u_i - \sum_j \nabla_j h_{ij}(u) - \beta_i(u) = v_i(x) \quad , \tag{14.1}$$

with h_{ij}, β, and v_i specified below. As a comparison function throughout this section we use

$$\mu(x) \equiv (1+|x|^2)^{1/2} \sim \begin{cases} 1 , & |x| \to 0 \\ |x| , & |x| \to \infty \end{cases} \quad . \tag{14.2}$$

We restrict our attention to $\mathbb{R}^3$ for clarity. Basic generalizations to $\mathbb{R}^d$ exist, $d \geq 3$.

PROPOSITION 14.1. *Let* u *be a smooth solution to* (14.1), *such that for* $c_1 < \infty$,

$$|u(x)| \leq c_1 \mu(x)^{-1} \tag{14.3}$$

and

$$u(x) \in L_2(\mathbb{R}^3) \quad . \tag{14.4}$$

Suppose $h_{ij}(u)$ *and* $\beta(u)$ *are smooth functions and that there exists* $\varepsilon > 0$ *such that for* $|u(x)|$ *sufficiently small, there exists a constant* $z_0 < \infty$ *such that for all* i,j,

$$|h(u)| \leq z_0 |u|^{5/3+\varepsilon} \tag{14.5}$$

$$|\beta(u)| \leq z_0 |u|^{8/3+\varepsilon} \quad . \tag{14.6}$$

Furthermore suppose

$$|v(x)| \leq c_1 \mu(x)^{-4} \tag{14.7}$$

and

$$\mu(x)|v(x)| \in L_1(\mathbb{R}^3) \quad . \tag{14.8}$$

Then there exists $c < \infty$ *such that*

$$|u(x)| \leq c\mu(x)^{-2} \quad . \tag{14.9}$$

Remark. The assumption $u \in L_2(\mathbb{R}^3)$ of (14.4) can be relaxed, at the expense of exponents in (14.5-6). For the application in §15, we use $p = 2$, so we restrict attention to that case.

The first step in the proof is an intermediate bound required to start the iteration.

PROPOSITION 14.2. *Let* $p < 3/2$. *There exists a constant* $c(p)$ *such that*

$$|u(x)| \leq c(p)\mu(x)^{-p} \quad . \tag{14.10}$$

Proof of Proposition 14.1, assuming Proposition 14.2. Start by rewriting (14.1) using the Green's function $G(x-y) = (4\pi|x-y|)^{-1}$ and $H_i(x-y) = \nabla_i G(x-y)$ as

$$-u_i(x) = (G(v_i+\beta_i))(x) + \sum_j (H_j, h_{ij})(x) \quad . \tag{14.11}$$

Note that $|u| < c(p)\mu^{-p}$ for $p < 3/2$. For h, β satisfying (14.5-6), by choosing p sufficiently close to $3/2$, there exists $\delta > 0$ and $c < \infty$ such that

$$|h(u(x))| \leq c\mu^{-(5/2+2\delta)} \in L_{6/5} \quad , \tag{14.12}$$

$$|\beta(u(x))| \leq c\mu^{-(4+2\delta)} \in L_1 \quad . \tag{14.13}$$

We now combine these results with the convolution bounds proved in Chapter VI, §4. By Proposition VI.4.4, with a new $c < \infty$,

$$|Hh(u(x))| \leq c\mu^{-(3/2+\delta)} \in L_2 \quad . \tag{14.14}$$

By assumption $u_i \in L_2$, and hence by (14.11), $G(v_i+\beta_i) \in L_2$. It follows by Proposition VI.4.7 that

$$\int_{\mathbb{R}^3} (v+\beta) = 0 \quad . \tag{14.15}$$

We wish to estimate $G(v_i - \beta_i)$ using Proposition VI.4.6, with $r = 0$. In fact hypotheses (a), (c) have been verified, while (14.8) ensures (b). Thus

$$|G(v_i + \beta_i)| \leq c\mu^{-2} \quad . \tag{14.16}$$

Inserting (14.14), (14.16) in (14.11) yields

$$\begin{aligned} |u(x)| &\leq c_2\mu^{-2} + c_3\mu^{-(3/2+\delta)} \\ &\leq c_4\mu^{-(3/2+\delta)} \quad , \end{aligned} \tag{14.17}$$

assuming $\delta \leq 1/2$. Hence we obtain a decay with exponent greater than 3/2, and hence better than (14.10). In fact, the improvement in the exponent from (14.10) to (14.17) is greater than δ.

If we start now with (14.17) and redo the bounds (14.12-17), we obtain

$$|h(u(x))| \leq c\mu^{-5/2(1+\delta)}$$

and by Proposition VI.4.4,

$$|Hh(u(x))| \leq c\mu^{-(3/2+5\delta/2)} \quad ,$$

yielding

$$|u(x)| \leq c_2\mu^{-2} + c_5\mu^{-(3/2+5\delta/2)} \leq c_6\mu^{-(3/2+5\delta/2)} \quad ,$$

assuming $5\delta/2 \leq 1/2$. Thus we improved from exponent $3/2+\delta$ to $3/2+5\delta/2$. Thus after a finite number j of steps, $(2.5)^j\delta > 1/2$, and we obtain

$$|u(x)| \leq c\mu^{-2} \quad ,$$

as desired.

Now we return to Proposition 14.2. Let us choose a cutoff function $b_R(x) = b(x/R)$, where as usual $b \in C_0^\infty$, $0 \leq b \leq 1$, and b is monotonic decreasing in $|x|$, satisfying

$$b(x) = \begin{cases} 1, & |x| \leq 1 \\ 0, & |x| \geq 2 \end{cases} \quad . \tag{14.18}$$

Define

$$\mu_R(x) = b_R(x)^2 \mu(x) \quad . \tag{14.19}$$

Also let

$$S_R = \sup_x |\mu_R(x)^p u(x)| \quad . \tag{14.20}$$

Since μ_R has compact support and u is assumed smooth, $S_R < \infty$. We are interested in bounds on S_R which are uniform in R. Note S_R is monotonic increasing in R (for large R).

PROPOSITION 14.3. *Under the hypotheses of Proposition* 14.1, *and given* $1 \leq p < 3/2$, *there exist constants* $\delta > 0$ *and* $a_0, a_1 < \infty$, *all independent of* R, *such that*

$$S_R \leq a_0 + a_1 S_R^{1-\delta} \quad . \tag{14.21}$$

COROLLARY 14.4. *The bound* (14.10) *of Proposition* 14.2 *holds.*

Proof. If $S_R \leq 1$ for all R, then $|u|$ is bounded by μ^{-1-p}. If not, for $R > \bar{R}$, $S_R \geq 1$. Dividing by $S_R^{1-\varepsilon}$, (14.21) then yields $S_R^\varepsilon \leq a_0 + a_1$, and S_R is bounded by $(a_0 + a_1)^{1/\varepsilon} \equiv c(p)$, as desired.

LEMMA 14.5. *There exists a constant* $M < \infty$, *independent of* R, *such that for* $p > 1$,

$$|\nabla \mu_R^p| \leq M \mu_R^{p-1} \quad , \tag{14.22}$$

and

$$|\Delta\mu_R^p| \leq M\mu_R^{p-1}\mu^{-1} \quad . \tag{14.23}$$

Proof. Note the bounds obtained from differentiation:

$$|\nabla\mu^p| \leq p\mu^{p-1} \quad , \tag{14.24}$$

$$|\Delta\mu^p| \leq p(p+1)\mu^{p-2} \quad . \tag{14.25}$$

For $R \geq 1$, and $2p > 1$,

$$\nabla b_R^{2p} = 2pb_R^{2p-1}\nabla b_R = 2pb_R^{2p-1}R^{-1}(\nabla b)(x/R) \quad . \tag{14.26}$$

Multiply by $1 = \mu\mu^{-1}$, and note $\mu \leq 3R$ on the support of b_R. Also note $|\nabla b| \leq$ const. Hence for $2p > 1$:

$$|\nabla b_R^{2p}| \leq \text{const } b_R^{2p-1}\mu^{-1} \quad . \tag{14.27}$$

Now the bound (14.22) follows from $b_R \leq 1$:

$$\begin{aligned} |\nabla(\mu^p b_R^{2p})| &\leq \text{const}(\mu^{p-1}b_R^{2p} + \mu^{p-1}b_R^{2p-1}) \\ &\leq \text{const } \mu_R^{p-1} \quad , \end{aligned} \tag{14.28}$$

as claimed. For the bound (14.23), use

$$\Delta(\mu^p b_R^{2p}) = (\Delta\mu^p)b_R^{2p} + 2(\nabla\mu^p)\cdot(\nabla b_R^{2p}) + \mu^p(\Delta b_R^{2p}) \quad . \tag{14.29}$$

Thus with $p > 1$, we infer from (14.24-27) and the derivative of (14.26), that

$$\begin{aligned} |\Delta(\mu^p b_R^{2p})| &\leq \text{const}(\mu^{p-2}b_R^{2p} + \mu^{p-1}b_R^{2p-1} + \mu^{p-2}b_R^{2p-2}) \\ &\leq \text{const } \mu_R^{p-1}\mu^{-1} \quad . \end{aligned}$$

LEMMA 14.6. *Under the hypotheses of Proposition* 14.1, *and for given* $p > 1$, *there is a constant* M *independent of* $R > 1$ *such that*

$$\mu_R^p|u(x)| \leq M \int G(x-y)(\mu_R^p|\beta| + \mu_R^p|v| + \mu_R^{p-1}|h| + \mu_R^{p-1}\mu^{-1}|u|)(y)\, dy$$

$$+ M \int |H(x-y)|(\mu_R^p|h(y)| + \mu_R^{p-1}|u(y)|)\, dy \quad . \tag{14.30}$$

Proof. We use (14.1) to derive an equation for $\mu_R^p u$,

$$\Delta(\mu_R^p u_i) - \sum_j \nabla_j(\mu_R^p h_{ij}) - \mu_R^p(\beta + v)$$

$$= \sum_j [-h_{ij}(\nabla_j \mu_R^p) + 2\nabla_j(u_i \nabla_j \mu_R^p)] - u_i \Delta\mu_R^p \quad . \tag{14.31}$$

Integrating with the Green's function yields

$$-\mu_R^p u = \int G(x-y)\left\{\mu_R^p(\beta + v) - \sum_j h_{ij}\nabla_j \mu_R^p - u\Delta\mu_R^p\right\} dy$$

$$+ \int H_j(x-y)\left\{2u\nabla_j \mu_R^p + \sum_j \mu_R^p h_{ij}\right\} dy \quad . \tag{14.32}$$

Inserting absolute values and using Lemma 14.5 yields (14.30).

Proof of Proposition 14.3. Break the integration in (14.30) into two regions,

$$< \equiv \{y: |x-y| < 1\}$$

$$> \equiv \{y: |x-y| > 1\} \quad .$$

Write (14.30) as $\mu_R^p|u(x)| \leq \int_< + \int_>$. By assumption, $\mu|u|$ is bounded. Hence by (14.5-6),

$$|h| \leq \text{const}\ \mu^{-5/3} \quad , \qquad |\beta| \leq \text{const}\ \mu^{-8/3} \quad . \tag{14.33}$$

Thus

$$\mu^{p-1}|u| + \mu^{p}(|h| + |\beta| + |v|) \leq \text{const} \quad . \tag{14.34}$$

Hence, since G and H are locally integrable,

$$\int_{<} \leq \text{const} \int_{<} (G(x-y) + |H(x-y)|)\, dy \leq \text{const} \quad ,$$

with the constant independent of R.

Next we bound $\int_{>}$, uniformly in R. Note the integrand is compactly supported, so convergence is not a problem. We consider the separate terms. Using (14.7) we bound the v-term,

$$\begin{aligned} \int_{>} \mu_R^p(y)\,|v(y)|\,G(x-y)\, dy &\leq c_1 \int G(x-y)\mu(y)^{-4+p}\, dy \\ &\leq \text{const} \quad , \end{aligned}$$

because $p < 3/2$. Again the constant is R-independent.

Five terms remain to be estimated. We divide these bounds into three lemmas.

LEMMA 10.7. *The* u *terms in* $\int_{>}$ *satisfy* (14.21).

Proof. By the Schwarz inequality,

$$\int_{>} G(x-y)\mu_R^{p-1}(y)\mu(y)^{-1}|u(y)|\, dy \leq \text{const}\, \|u\|_{L_2} \left(\int_{>} \mu(x-y)^{-2}\mu(y)^{-4+2p} dy \right)^{1/2} . \tag{14.35}$$

Since $p < 3/2$, $-4 + 2p < -1$. Thus (14.35) is bounded by $\text{const}\|u\|_{L_2} \leq$ const, using assumption (14.4). The other u term is

$$\int_> |H(x-y)|\mu_R^{p-1}|u(y)|\ dy \le \int_> \mu(x-y)^{-2}\mu_R^{p-1}(y)|u(y)|\ dy$$

$$\le \sup_y \left|\mu_R^p\, u(y)\right|^{\frac{p-1}{p}} \int_> \mu(x-y)^{-2}|u|^{1/p}\ dy$$

$$\le S_R^{1/p'}\ \|u\|_{L_2}^{1/p} \left(\int_> \frac{1}{\mu(x-y)^{4p/(2p-1)}}\right)^{(2p-1)/2p} \tag{14.36}$$

In the last step we use Hölder's inequality. The restriction $p < 3/2$ means

$$3 < \left(\frac{4p}{2p-1}\right) \quad ,$$

and therefore (14.36) is convergent. Hence using (14.4), the contribution (14.36) is bounded by

$$\text{const}\ S_R^{1/p'} = \text{const}\ S_R^{1-1/p} \quad , \tag{14.37}$$

as desired in (14.21).

LEMMA 14.8. *The* v *term in* $\int_>$ *is bounded by a constant.*

<u>Proof</u>. Since $\mu^4 v \in L_\infty$ by assumption, and $p < 3/2$, there exists $0 < \delta$ sufficiently small so that $p + \delta < 3/2$. Then

$$\int_> G(x-y)\mu^p|v(y)|\ dy \le \left(\int_> G(x-y)\mu^{-2-\delta}\ dy\right) \sup|\mu^{2+p+\delta}v|$$
$$\le \text{const} \quad ,$$

to complete the proof.

LEMMA 14.9. *The* β *and* h *terms in* $\int_>$ *satisfy* (14.21).

Proof. Hence we use the assumptions (14.5-6). Note in the integrands $G(x-y) \leq \text{const}\ \mu(x-y)^{-1}$, $|H(x-y)| \leq \text{const}|x-y|^{-2} \leq \text{const}\ \mu(x-y)^{-2}$, so we must estimate

$$\begin{aligned}
\mu_R^p|\beta| + \mu_R^{p-1}|h| &\leq \text{const}\ \mu_R^p|u|^{8/3+\varepsilon} + \mu_R^{p-1}|u|^{5/3+\varepsilon} \\
&\leq \text{const}\ \mu_R^{p-1}|u|^{5/3+\varepsilon} \\
&= \text{const}(\mu_R^p|u|)^{(p-1)/p}|u|^{5/3-(p-1)/p+\varepsilon} \\
&\leq \text{const}\ S_R^{1/p'}|u|^{5/3-1/p'+\varepsilon} .
\end{aligned}$$

Here we used $|\mu_R u| \leq |\mu u| \leq \text{const}$ in the second inequality. Thus by Hölder's inequality

$$\begin{aligned}
&\int_{>} G(x-y)(\mu_R^p|\beta| + \mu_R^{p-1}|h|)\,dy \\
&\qquad \leq \text{const}\ S_R^{1/p'}\int_{>} \frac{dy}{|x-y|}\ |u|^{5/3-1/p'+\varepsilon} \\
&\qquad \leq \text{const}\ S_R^{1/p'}\|u\|_{L_2}^{2/\alpha'}\left(\int_{>} \frac{dy}{|x-y|^\alpha}\right)^{1/\alpha}
\end{aligned} \tag{14.38}$$

where $\alpha = [2(5/3 - 1/p' + \varepsilon)^{-1}]'$. We require $\alpha > 3$ (i.e., $\alpha' < 3/2$) to ensure a finite bound. But $p < 3/2$ so $p' > 3$. Hence

$$\alpha' < 2\left(\frac{5}{3} - \frac{1}{3} + \varepsilon\right)^{-1} < \frac{3}{2} ,$$

ensuring that (14.38) is bounded by

$$\text{const}\ S_R^{1/p'} = \text{const}\ S_R^{1-1/p} .$$

The final term is the integral of

$$|H(x-y)|\mu_R^p|h(y)| \leq \text{const } \mu(x-y)^{-2}\mu_R^p|u|^{5/3+\varepsilon}$$

$$= \text{const } \mu(x-y)^{-2}(\mu_R^{p-2\delta}|u|^{1-\delta})(|u|^{2/3+\varepsilon+\delta}\mu_R^{2\delta}) \quad .$$

Choose $\delta < \varepsilon/2$. Note $(p-2\delta)/(1-\delta) < p$. Thus

$$|H(x-y)|\mu_R^p|h(y)| \leq \text{const } S_R^{1-\delta}\mu(x-y)^{-2}|u|^{2/3+\varepsilon/2} \quad .$$

By Hölder's inequality

$$\int_{>}|H(x-y)|\mu_R^p|h(y)| \; dy \leq \text{const } S_R^{1-\delta}\, \|u\|_{L_2}^{2/3+\varepsilon/2}\left(\int_{>}\frac{1}{\mu(x-y)^{2\alpha}}\right)^{1/\alpha} \quad .$$

Here $\alpha' = 2(2/3+\varepsilon/2)^{-1}$. Thus $\alpha' < 3$, and $\alpha > 3/2$. Hence $2\alpha > 3$ and the bound is finite, to complete the proof.

IV.15. THE DECAY OF $|(F_A)_L|$

We prove that as $|x| \to \infty$, (Φ,F) is $O(|x|^{-2})$. This comes in the process of completing the proofs of Theorems 10.2 and 10.5. When $\lambda > 0$, the proof of power law decay is an application of the results in Chapter VI.4. When $\lambda = 0$, as a first step, we prove that $|f_L|$, $|g|_L$ are $O(|x|^{-1})$. We then apply the method developed in the previous section. The basic result is the following

PROPOSITION 15.1. *Let* (A,Φ) *be a smooth, finite action solution to equations* (1.4).

CASE 1: *If* $\lambda > 0$, *then the estimate of Theorem* 10.2c *holds.*
CASE 2: *Suppose* $\lambda = 0$ *and there exists* $c_1 < \infty$ *such that*

$$|f| + |g| \leq c_1(1+|x|)^{-1} \quad . \tag{15.1}$$

Then there exists $c < \infty$ *such that*

$$|f| + |g| \leq c(1+|x|)^{-2} \quad . \tag{15.2}$$

Proof. For $\lambda > 0$ we have from (9.17) that

$$-\Delta(\Phi,f) = \lambda w(\Phi,f) - 2(g_j, \nabla_{A_j} f) + (\Phi, P_f \Lambda(\Psi)) \quad , \tag{15.3}$$

where

$$P_f = \begin{pmatrix} I & 0 \\ 0 & 0 \end{pmatrix} \quad .$$

Each $(\Phi, f_i)_{i=1}^{3} \in L_2$ and decays uniformly to zero as $|x| \to \infty$ (Proposition 11.1). Further, it has been established (Propositions 12.5, 13.3-4) that each term on the right hand side of equation (15.3) decays to zero exponentially as $|x| \to \infty$. Therefore, Case 1 follows from (15.3) and Propositions VI.4.6-7.

As for Case 2, write $\Psi = (f_i, g_i)$ and $u = (u_1, \ldots, u_6) = ((\Phi, f_1), \ldots, (\Phi, g_3))$. Due to Theorem 10.3, there exists $r < \infty$ such that

$$\inf_{|x|>r} |\Phi(x)| > \frac{1}{2} \quad . \tag{15.4}$$

We use this fact to control terms in (9.17). Equation (9.17) has the form

$$-\Delta u_i + \nabla_j h_{ij}(u) = v_i \quad , \tag{15.5}$$

where

$$h_{ij}(u) = 2(1-b_r)|\Phi|^{-2} (P_g)_{jk} u_k u_i \quad ,$$

$$P_g = \begin{pmatrix} 0 & 0 \\ 0 & I \end{pmatrix}$$

and

$$v_i = -2\nabla_j (b_r (g_j, \Psi_i)) - 2\nabla_j ((1-b_r)(g_{jT}, \Psi_i) + (\Phi, \Lambda(\Psi)_i) \quad . \tag{15.6}$$

Because of the choice of r,

$$|h| \leq 8\,|u|^2 \quad . \tag{15.7}$$

As for v_i, the first term is compactly supported and the last term is exponentially decaying to zero as $|x| \to \infty$ (see (12.3) and Proposition 12.5). Expanding out the second term we obtain

$$|\nabla_j((1-b_r)(g_{jT},\Psi)| \leq |g_{jT}|(|\nabla_{A_j}\Psi| + |\Psi||\nabla_j b_r|) + 12|\Psi||g||g_T|. \tag{15.8}$$

Here we have used (9.2b). The function $|\nabla_A \Psi|$ is bounded (see Proposition 12.4). Hence this term decays exponentially to zero also.

These last remarks have established that (15.5) satisfies the conditions of Proposition 14.1; the conclusion of Proposition 14.1 proves the claim for Case 2.

COROLLARY 15.2. *Let* (A,Φ) *be a smooth, finite action solution to equations* (1.4) *with* $\lambda = 0$. *Suppose that* $\lim_{|x|\to\infty} |\Phi| \to 1$ *(cf. Theorem* 10.3*). Then Statement* (c) *of Theorem* 10.5 *holds.*

<u>Proof.</u> From Proposition 15.1, $|g|^2$ is $O(|x|^{-4})$. Now use Proposition VI.4.3 and equation (9.4) with $\lambda = 0$.

We will complete the proof of Theorem 10.5 by establishing that (15.1) holds. This is stated in the following proposition.

PROPOSITION 15.3. *Let* (A,Φ) *be a smooth, finite action solution to equations* (1.4) *with* $\lambda = 0$. *Then there exists* $c_1 < \infty$ *such that* $|f|$, $|g|$ *satisfy* (15.1).

<u>Proof.</u> From (III.6.21), (9.6) and (9.7),

$$|\Psi|\Delta|\Psi| \geq (\Psi,\nabla_A^2\Psi) \geq -\,|\Psi||\Lambda(\Psi)| \qquad \text{a.e.} \tag{15.9}$$

As we remarked previously $|\Lambda_1(\Psi)| \sim O(e^{-|x|})$. Define a comparison function

$$s(x) = \frac{1}{4\pi} \int_{\mathbb{R}^3} dy \frac{1}{|x-y|} |\Lambda(\Psi)(y)| . \tag{15.10}$$

From Corollary VI.4.2,

$$-\Delta s = |\Lambda(\Psi)| , \tag{15.11}$$

and by Proposition VI.4.3, there exists $c_1 < \infty$ such that

$$s(x) \leq c_1 (1 + |x|)^{-1} . \tag{15.12}$$

From equations (15.9) and (15.12),

$$|\Psi| \Delta (s - |\Psi|) \leq 0 \qquad \text{a.e.} \tag{15.13}$$

The trick used to prove Proposition III.7.2 works as well here. Since $s > 0$, the function $|\Psi|$ is C^∞ in the support of $(s - |\Psi|)_-$. Further, $s - |\Psi| = 0$ on the boundary of $\mathrm{supp}((s - |\Psi|)_-)$. By taking a slightly larger open set V we can ensure that $\partial \bar{V}$ is smooth, $0 < |\Psi| \in C^\infty(\bar{V})$ and $s - |\Psi| \geq 0$ on $\partial \bar{V}$. In V, we divide both sides of (15.3) by $|\Psi|$ and use the maximum principle to infer that

$$|\Psi| \leq s , \qquad x \in V . \tag{15.14}$$

But this last statement implies that $|\Psi| \leq s$ everywhere. Proposition 15.3 now follows from (15.12).

IV.16. THE ZEROS OF Φ

As was noted in §1, the set $Z(\Phi) = \{x \in \mathbb{R}^3 | \Phi(x) = 0\}$ may be important in classifying solutions to (1.2-3). We prove properties of Z which we list in the following propositions.

In this section, the gauge group G is any simple Lie group. The first result is an immediate consequence of Chapter V (see Theorem V.1.1).

PROPOSITION 16.1. *Let* (A,Φ) *be a smooth solution to* (1.4) *with* $\lambda \geq 0$ *and* G *any simple Lie group. Then* $Z(\Phi)$ *is a real analytic variety.*

Proposition 16.1 immediately rules out the existence of smooth solutions to the equations in which Z is a line segment with endpoints. Theorems 10.1, 10.3 of this chapter rule out the possibility that Z is a line in $\mathbb{R}^3$. (These two theorems are stated for $G = SU(2)$ but it is not hard to see that they generalize to arbitrary G.) Proposition 16.1 also rules out sets Z with nonzero Lebesgue measure in $\mathbb{R}^3$.

PROPOSITION 16.2. *Let* (A,Φ) *be a smooth, finite action solution to* (1.4b) *with* $\lambda = 0$. *It is not necessary to assume* (1.4a). *Suppose that* $\lim_{|x|\to\infty} |\Phi| = 1$ *uniformly with* $|x|$. *Then* $Z(\Phi)$ *cannot be the boundary of an open set in* $\mathbb{R}^3$.

We conjecture that for $\lambda = 0$, Propositions 16.1 and 16.2 rule out codimension 1 components of $Z(\Phi)$ completely.

Concerning codimension 2 components, Proposition 16.1 allows for the existence of components of Z which are real analytic closed curves in $\mathbb{R}^3$ of bounded extent.

CONJECTURE. *Let* (A,Φ) *be smooth with* $\lim_{|x|\to\infty} |\Phi| = 1$. *Suppose that* (A,Φ) *is a solution to* (1.4b) *with* G *any simple Lie group and* $\lambda > 0$. *Do not assume* (1.4a). *Then the set* $Z(\Phi)$ *is a finite set of points.*

Our next proposition gives credence to this conjecture. Let L be a bounded, piecewise smooth curve in $\mathbb{R}^3$. We define the diameter of L to be the number

$$\ell = \sup_{(x,y)\in L} |x-y| \quad . \tag{16.1}$$

PROPOSITION 16.3. *Let* (A,Φ) *be smooth with* $\lim_{|x|\to\infty} |\Phi| = 1$. *Assume that* (1.4b) *is satisfied for* $\lambda \geq 0$. *There exists a polynomial* $P(t)$ *such that the size* $|\gamma|$ *of each piecewise smooth curve* $\gamma \in Z(\Phi)$

is bounded by $P(\mathscr{A})$. *Here* $\mathscr{A}(A,\Phi)$ *is the action.*

Given a smooth section Φ of $\mathbb{R}^3 \times g$, we define a unit length section over $\mathbb{R}^2 \setminus Z(\Phi)$

$$\hat{\Phi} = \Phi|\Phi|^{-1} \quad . \tag{16.2}$$

DEFINITION 16.4. *The singular set* $S(\Phi)$ *is the complement of the set in* $\mathbb{R}^3$ *on which* $|\nabla\hat{\Phi}| \in L_\infty$.

Although $|\nabla\hat{\Phi}|$ is not gauge invariant, the set $S(\Phi)$ is left invariant by C^1 gauge transformations.

PROPOSITION 16.5. *Let* (A,Φ) *be as in Proposition* 16.2. *Then*

$$Z(\Phi) = S(\Phi) \quad . \tag{16.3}$$

The proofs of these propositions comprise the remainder of this section.

Proof of Proposition 16.2. The function $w = \frac{1}{2}(1-|\Phi|^2)$ satisfies (9.4) with $\lambda = 0$. Suppose $\partial\bar{V} \subset Z(\Phi)$ for some open, bounded set $V \in \mathbb{R}^3$. From the maximum principle (Chapter VI.3), w cannot have a minimum in V. Hence $w \equiv 1/2$ in V and $V \subset Z(\Phi)$. By Proposition 16.1, this is impossible.

Proof of Proposition 16.3. Let $L \in Z(\Phi)$ be a piecewise continuous curve of diameter ℓ. We prove Proposition 16.3 by showing that there is a tubular neighborhood enclosing L of minimum radius ρ and length ℓ in which $w > 1/4$. The number ρ will turn out to depend polynomially on $\mathscr{A}(A,\Phi)$. We then show that the action gives an upper bound to the volume of the set $w > 1/4$. Because the volume of this set is proportional to ℓ, we get an upper bound on ℓ.

We begin by remarking that $|\Phi| < 1$ by the maximum principle applied to (9.4). The knowledge that $|\Phi|$ is bounded allows us to apply Proposition V.8.1 to bound $\||g|\|_{H_1}$ by $\mathscr{A}$. Specifically,

$$\| |g| \|_{H_1} \leq c_1(\mathscr{A} + \mathscr{A}^3) \quad , \tag{16.4}$$

with $c_1 < \infty$, independent of (A,Φ). By a Sobolev inequality, $|g| \in L_p$, $\forall$ $p \in [2,6]$ (Proposition VI.2.4). Furthermore, because

$$|dw| = |(\Phi,g)| \leq |g| \quad , \tag{16.5}$$

w is Hölder continuous with exponent $1/2$ (Corollary VI.2.7). This means that for any $x,y \in \mathbb{R}^3$,

$$|w(x) - w(y)| \leq c_2|x-y|^{1/2}\|g\|_{L_6} \leq c_2'|x-y|^{1/2}\| |g| \|_{H_1} \quad , \tag{16.6}$$

with c_2, $c_2' < \infty$, independent of (A,Φ). Equation (16.6) implies that $w(y)$ is uniformly close to $w(x)$ for y close to x. By the triangle inequality,

$$w(y) \geq w(x) - c_2'|x-y|^{1/2} \| |g| \|_{H_1} \quad . \tag{16.7}$$

Equation (16.7) implies the following lemma:

LEMMA 16.6. *If* $w(x) = 1/2$, *then* $w(y) > 1/4$ *whenever*

$$|x-y| \leq \frac{1}{16} \left(c_2' \| |g| \|_{H_1}\right)^{-2} \equiv \rho \quad . \tag{16.8}$$

Clearly the set $V = \{x \in \mathbb{R}^3 | \, w(x) > 1/4\}$ has Lebesgue measure

$$\mu = \int_V dx \geq \pi\rho^2\ell \quad . \tag{16.9}$$

In order to use (16.9) to bound ℓ we use the Green's function representation for w. The function w satisfies the differential inequality

$$\Delta w = -|g|^2 + \lambda(1-2w)w \geq -|g|^2 \quad . \tag{16.10}$$

Define a function $s(x)$ by

$$s(x) = \frac{1}{4\pi} \int dy \frac{|g(y)|^2}{|x-y|} . \tag{16.11}$$

It follows from Propositions VI.4.1-2, that $s(x)$ is smooth, decays uniformly to zero as $|x| \to \infty$ and satisfies $\Delta s = -|g|^2$. By the maximum principle,

$$s(x) \geq w(x) , \tag{16.12}$$

with equality if $\lambda = 0$. Therefore

$$\frac{1}{4} \int_V dx \equiv \frac{1}{4} \mu \leq \int_V w dx \leq \int_V s dx = \int_V dx \int_{\mathbb{R}^3} dy \frac{|g(y)|^2}{|x-y|} \tag{16.13}$$

We interchange the order of integration in (16.13) and evaluate $\int_V dx\ |x-y|^{-1}$. In fact, for $r \geq 0$,

$$\int_V \frac{dx}{|x-y|} = \int_{V \cap \{|x-y|<r\}} \frac{dx}{|x-y|} + \int_{V \cap \{|x-y| \geq r\}} \frac{dx}{|x-y|}$$

$$\int_V \frac{dx}{|x-y|} \leq 2\pi r^2 + \frac{\mu}{r} . \tag{16.14}$$

Equation (16.14) is valid for any $r \geq 0$. Minimizing right hand side of (16.14) over r we obtain

$$\int_V \frac{dx}{|x-y|} \leq (2\pi)^{1/3} \mu^{2/3} . \tag{16.15}$$

Together, (16.13) and (16.15) imply

$$\frac{1}{4} \mu \leq (2\pi)^{1/3} \mu^{2/3} \|g\|_{L_2}^2 . \tag{16.16}$$

This last equation establishes an upper bound for μ

$$\mu \leq 2\pi \cdot 64 \|g\|_{L_2}^6 \quad ,$$

and therefore an upper bound for ℓ:

$$\ell \leq 128\rho^{-2}\|g\|_{L_2}^6 \;=\; 8\cdot(16)^3(c_2')^2\||g|\|_{H_1}^4 \|g\|_{L_2}^6 \quad . \tag{16.17}$$

Proposition 16.3 follows by substituting $\mathscr{A}$ for $\|g\|_{L_2}^2$ and using (16.4).

Proof of Proposition 16.5. By assumption (A,Φ) is smooth so $\hat{\Phi}$ is smooth where $|\Phi| > 0$; that is $S(\Phi) \subset Z(\Phi)$. We use the maximum principle to prove the converse. Let V be an open bounded set with smooth boundary $\partial\bar{V}$ such that $|\Phi| \geq \delta > 0$ on $\partial\bar{V}$. Assume that $|\nabla\hat{\Phi}| \in L_\infty(\bar{V})$. We show that $|\Phi|$ does not vanish in V. Let $\eta \in L_2^1(V)$ such that $\eta = 0$ on $\partial\bar{V}$ a.e., but otherwise arbitrary. Then

$$\int_V (D_A\eta, D_A\Phi) \;=\; 0 \quad . \tag{16.18}$$

Set $\eta = \psi\hat{\Phi}$ for $\psi \in L_2^1(V)$, vanishing a.e. on $\partial\bar{V}$. This choice of $\eta \in L_2^1(V;g)$ since $\hat{\Phi}$, $\nabla\hat{\Phi} \in L_\infty(\bar{V})$. Inserting into (16.18) the above choice for η, we obtain

$$\int_V dx(\nabla_j\psi\nabla_j|\Phi| + \psi|\Phi|\alpha^2) \;=\; 0 \quad , \tag{16.19}$$

which is valid for all compactly supported ψ. The function

$$\alpha^2 \;=\; (\nabla_A\hat{\Phi},\nabla_A\hat{\Phi}) \tag{16.20}$$

is bounded on V. Let M^2 be its maximum. Let h be the unique C^∞ solution to the equation

$$\begin{aligned} -\Delta h + M^2 h &= 0 && \text{in } V \\ h &= \delta && \text{on } \partial\bar{V} \end{aligned} \tag{16.21}$$

The strong maximum principle (Chapter VI.3) implies that $h > 0$ in V. The weak maximum principle implies that $|\Phi| \geq h$ in V. Thus $Z(\Phi) \cap \mathbb{R}^3 \setminus S(\Phi) = \emptyset$ and $Z(\Phi) \subset S(\Phi)$. Equality follows as claimed.

IV.17. THE HIGGS FIELD AS AN ORDER PARAMETER

We answer in the affirmative the question of whether the Higgs field acts as a local order parameter by the $1 - |\Phi(x)|^2$ bounds for $|F_A(x)|^2$ and $|D_A\Phi(x)|^2$. We consider only the case $\lambda = 0$. Our first result concerns $g = D_A\Phi$.

PROPOSITION 17.1. *Let* (A,Φ) *be a smooth, finite action solution to* (1.4) *with* $\lambda = 0$. *Take* G *to be any simple Lie group. Suppose that* $\lim_{|x|\to\infty} |\Phi|^2 = 1$. *Then there exists the pointwise bound*

$$|g(x)|^2 \leq 4\sqrt{2}\ \|f\|_\infty (1-|\Phi(x)|^2) \quad , \tag{17.1}$$

where

$$\|f\|_\infty = \sup_{x\in\mathbb{R}^3} |F_A(x)|$$

is bounded by a polynomial in $\mathscr{A}(A,\Phi)$ *(see Proposition* 10.6*).*

Concerning a pointwise bound on $f = *F_A$ we have,

PROPOSITION 17.2. *Let* (A,Φ) *be as in the previous proposition. Restrict the gauge group* $G = SU(2)$. *Then there exists the pointwise bound*

$$|f(x)|^2 + 12\sqrt{2}\|f\|_\infty |g(x)|^2 \leq (194\|f\|_\infty^2 + 2\sqrt{2}\|f\|_\infty)(1 - |\Phi(x)|^2) \ . \tag{17.2}$$

The bounds given above can be sharpened if we assume that equation (1.14) is satisfied.

PROPOSITION 17.3. *Let* (A,Φ) *be as in Proposition* 13.1 *and satisfy either* (1.14a) *or* (1.14b). *Then for all* $x \in \mathbb{R}^3$

$$|g(x)| = |f(x)| \leq 2\sqrt{2}(1 - |\Phi(x)|^2) \quad . \tag{17.3}$$

We remark that all the results in this section follow from the maximum principle. The remainder of this section contains the proofs of the above propositions.

Proof of Proposition 17.1. Using the identity (9.19) and equation (9.13) we obtain

$$-\Delta \frac{|g|^2}{2} = -|\nabla_A g|^2 - |\Phi|^2 |g_T|^2 - 4\varepsilon^{ijk}(f_i, g_j g_k) \quad , \tag{17.4}$$

and hence the inequality

$$-\Delta \frac{|g|^2}{2} \leq 4\sqrt{2}|f||g|^2 \leq 4\sqrt{2}\|f\|_\infty |g|^2 \quad . \tag{17.5}$$

Recall that w satisfies equation (9.4) with $\lambda = 0$. Thus

$$\Delta(4\sqrt{2}\|f\|_\infty w - |g|^2/2) \leq 0 \quad . \tag{17.6}$$

Both w and $|g|^2$ decay uniformly to zero as $|x| \to \infty$. The maximum principle (Chapter VI.3) and (17.6) imply (17.1).

Proof of Proposition 17.2. Define a set $V_\rho \subset \mathbb{R}^3$, $1 \geq \rho \geq 0$ by

$$V_\rho = \{x \in \mathbb{R}^3 \mid |\Phi(x)|^2 > \rho\} \quad . \tag{17.7}$$

By assumption, equations (9.2a) and (9.3a) are satisfied. Taking the absolute value of both sides of (9.2a) and (9.3a) gives the following inequalities:

$$2|\nabla_A g|^2 \geq |\Phi|^2 |f_T|^2 \geq \rho |f_T|^2 \quad ,$$

and

$$2|\nabla_A f|^2 \geq |\Phi|^2 |g_T|^2 \geq \rho |g_T|^2 \quad , \quad x \in V_\rho \quad . \tag{17.8}$$

Using the first inequality above in (17.4) we obtain

$$-\Delta \frac{1}{2} |g|^2 \leq -\frac{\rho}{2} |f_T|^2 - \rho |g_T|^2 + 4\sqrt{2}|f||g|^2 , \qquad x \in V_\rho \tag{17.9}$$

An equation for $|f|^2$, similar to (17.4) can be derived from (9.10). It is

$$-\Delta \frac{1}{2} |f|^2 = -|\nabla_A f|^2 - |\Phi|^2 |f_T|^2 - \varepsilon^{ijk}(f_i, [g_j, g_k]) - \varepsilon^{ijk}(f_i, [f_j, f_k]). \tag{17.10}$$

The important fact is that the last term in (17.10) is a commutator. Hence it is quadratic in f_T and,

$$|\varepsilon^{ijk}(f_i, [f_j, f_k])| \leq 6\sqrt{2}|f||f_T|^2 \leq 6\sqrt{2}\|f\|_\infty |f_T|^2 . \tag{17.11}$$

Together, equations (17.8, 10, 11) imply the following inequality

$$-\Delta \frac{1}{2} |f|^2 \leq -\frac{\rho}{2} |g_T|^2 - \rho |f_T|^2 + 2\sqrt{2}\|f\|_\infty |g|^2 + 6\sqrt{2}\|f\|_\infty |f_T|^2 \tag{17.12}$$

which is valid for $x \in V_\rho$.

Multiply (17.9) by $12\sqrt{2}\rho^{-1}\|f\|_\infty$ and add it to (17.12). The resulting equation is

$$-\Delta \left(\frac{1}{2} |f|^2 + 3\sqrt{2}\rho^{-1}\|f\|_\infty |g|^2\right) \leq (96\rho^{-1}\|f\|_\infty^2 + 2\sqrt{2}\|f\|_\infty) |g|^2 , \quad x \in V_\rho . \tag{17.13}$$

On the boundary of V_ρ we have

$$\frac{1}{2} |f|^2 + 3\sqrt{2}\rho^{-1}\|f\|_\infty |g|^2 \Big|_{\partial \bar{V}_\rho} \leq \|f\|_\infty^2 \left(\frac{1}{2} + \rho^{-1} 24\right) . \tag{17.14}$$

Here we have used (17.1). Choose $\rho = 1/2$ and set

$$\mu = (194\|f\|_\infty^2 + 2\sqrt{2}\|f\|_\infty) . \tag{17.15}$$

Then adding μ times equation (9.4) to (17.13) we obtain,

$$-\Delta\left(\frac{1}{2}|f|^2 + 6\sqrt{2}\|f\|_\infty |g|^2 - \mu w\right) \leq 0 \quad , \quad x \in V_{1/2}$$

and

$$\frac{1}{2}|f|^2 + 6\sqrt{2}\|f\|_\infty |g|^2 - \mu w \leq 0 \quad , \quad x \in \partial\bar{V}_{1/2} \ . \qquad (17.16)$$

The maximum principle, applied to (17.16) in $V_{1/2}$ implies that

$$|f|^2 + 12\sqrt{2}\|f\|_\infty |g|^2 < \mu(1 - |\Phi|^2) \quad , \quad x \in V_{1/2} \ . \qquad (17.17)$$

However, because of the choice of μ, equation (17.17) is valid for all $x \in \mathbb{R}^3$, which proves the proposition.

<u>Proof of Proposition 17.3</u>. Starting from equation (9.10) for $\lambda = 0$ we have

$$-\nabla_A^2 f = [[f,\Phi],\Phi] - 4*(f \wedge f) \quad , \qquad (17.18)$$

where we have used the assumption that $g = \pm f$. Hence from Lemma III.7.1 and (17.18),

$$|f|\Delta|f| \geq -4\sqrt{2}|f|^3 \quad . \qquad (17.19)$$

Equation (9.4) reads

$$\Delta w = -|f|^2 \quad . \qquad (17.20)$$

We use w as a comparison function. From the two previous equations,

$$-|f|(\Delta(|f| - 4\sqrt{2}w) \leq 0 \quad . \qquad (17.21)$$

We conclude from the maximum principle (cf. the argument used in the proofs of Proposition III.7.2 and Proposition 15.3) that $|f| \leq 4\sqrt{2}w$ as claimed.

IV.18. MULTIMONOPOLES FOR SIMPLE GAUGE GROUPS

The existence of multimonopole solutions to (1.14) where the gauge group G is a compact, simple Lie group has also been established. These solutions, as when $G = SU(2)$, are generically not spherically symmetric. They can be interpreted as approximate superpositions of widely spaced, spherically symmetric monopoles.

As in the case where $\mathfrak{g} = \mathfrak{su}(2)$, the set of $(A,\Phi) \in C^{\infty}$ which satisfy the asymptotic conditions

$$\lim_{|x|\to\infty} |\Phi(x)| \to 1 \quad ,$$

$$\lim_{|x|\to\infty} |x|^{1+\delta}|D_A\Phi| = 0 \quad , \tag{18.1}$$

for some $\delta > 0$, separates into disjoint subsets. These subsets are labeled by homotopy classes of maps of S^2 as explained in Chapter II. Recall that if (A,Φ) satisfies (18.1) then a homotopy class $[(A,\Phi)] \in \pi_2(G/J)$ is defined. Here J is a Lie subgroup of G and it is constructed in the following way: Letting $\Phi(P)$ denote the asymptotic value of Φ on the z-axis,

$$J = \{g \in G | g^{-1}\Phi(P)g = \Phi(P)\} \quad . \tag{18.2}$$

That this construction makes sense is proved in Theorem II.3.1. As proved in Theorem II.3.1, the homotopy class $[(A,\Phi)] \in \pi_2(G/J)$ is specified by fixing $\ell-r$ integers, $\{n_a\}_{a=1}^{\ell-r}$. Here $\ell = \text{rank } G$ and $0 \leq r \leq \ell$.

The most general form for J is a direct product:

$$J = T' \times G' \quad , \tag{18.3}$$

where T' is a torus and G' is a simple Lie subgroup of G. The number $\ell-r$ is the dimension of the torus T'. This is all discussed in Chapter II.

A gross classification of the solutions to (1.14) is given by specifying the Lie subgroup J, or alternatively, $h = \Phi(P)$, and a set of integers $\{n_a\}_{a=1}^{\ell-r}$ that are compatible with the choice of J.

We define a set $\mathscr{H}(h,\{n_a\}_{a=1}^{\ell-r})$ which is indexed by a unit vector $h \in \mathfrak{g}$ and a set of integers $\{n_a\}_{a=1}^{\ell-r}$ as follows:

$$\mathscr{H}(h,\{n_a\}_{a=1}^{\ell-r}) = \Big\{ (A,\Phi) \in C^{\infty}(\mathbb{R}^3; T^* \otimes \mathfrak{g}) \oplus C^{\infty}(\mathbb{R}^3; \mathfrak{g}):$$

(a) $\mathscr{A}(A,\Phi) < \infty$.

(b) (A,Φ) satisfies (18.1).

(c) $\Phi(P) = h$.

(d) With J defined by (18.2), the class $[(A,\Phi)] \in \pi_2(G/J)$ is completely specified by $\{n_a\}_{a=1}^{\ell-r} \Big\}$.

If two vectors $h, h' \in \mathfrak{g}$ are conjugate, that is, if $h = g^{-1}h'g$ for some $g \in G$, then

$$\mathscr{H}(h,\{n_a\}_{a=1}^{\ell-r}) = g^{-1}\mathscr{H}(h',\{n_a\}_{a=1}^{\ell-r})g \quad . \tag{18.4}$$

Let $\mathscr{H}(h,\{n_a\}_{a=1}^{\ell-r})$ also denote the conjugacy class under the action of (18.4). This fact that $\mathscr{H}$ is a conjugacy class will be implicitly assumed.

THEOREM 18.1. *For every choice of* $\mathscr{H}(h,\{n_a\}_{a=1}^{\ell-r})$ *such that the set* $\{n_a\}_{a=1}^{\ell-r}$ *is nontrivial and nonpositive, there exists at least a countably infinite set of solutions to* (1.14a) *in* $\mathscr{H}(h,\{n_a\}_{a=1}^{\ell-r})$.

Remark. If the set $\{n_a\}_{a=1}^{\ell-r}$ is nontrivial and nonnegative the theorem is true for (1.14b) since the antimonopole solutions can be generated from the stated solutions by taking $\Phi \to -\Phi$.

In Theorem 18.2, we will be more specific about the structure of the solutions given by Theorem 18.1. Before stating Theroem 18.2, we review certain facts about the $O(3)$ symmetric solutions to equation (1.14a) [23,3,10].

The group of rotations, $O(3)$, acts on $\mathbb{R}^3$ in the usual way. An action of $O(3)$ on the vector bundle $(\mathbb{R}^3 \times T^* \otimes \mathfrak{g}) \oplus (\mathbb{R}^3 \otimes \mathfrak{g})$ can be defined by giving a lifting of the action of $O(3)$ to the principal bundle $\mathbb{R}^3 \times G$ which commutes with the natural projection $p: \mathbb{R}^3 \times G \to \mathbb{R}^3$. Such a lifting is uniquely defined by a homomorphism $L: \mathfrak{su}(2) \to \mathfrak{g}$ [5].

The Lie algebra $\mathfrak{su}(2)$ has a realization by the differential operators

$$\mathscr{L}_i = -\varepsilon^{ijk} x_j \frac{\partial}{\partial_x K} \tag{18.5}$$

on $C^\infty(\mathbb{R}^3)$, Let $L_i = L(\mathscr{L}_i)$. We assume that the homomorphism $L: \mathfrak{su}(2) \to \mathfrak{g}$ is not trivial. A configuration $(A,\Phi) \in C^\infty(\mathbb{R}^3; T^* \otimes \mathfrak{g}) \oplus C^\infty(\mathbb{R}^3; \mathfrak{g})$ is said to be $O(3)$-symmetric iff

$$(L_i + \mathscr{L}_i)(\Phi) = 0 \quad , \tag{18.6}$$

$$(L_i + \mathscr{L}_i)(A_j dx^j) = \varepsilon_{ijk} A_k dx^j \quad . \tag{18.7}$$

For given homomorphism $L: \mathfrak{su}(2) \to \mathfrak{g}$ and set $\mathscr{H}(h, \{n_a\}_{a=1}^{\ell-r})$, the complete set $(A,\Phi) \in \mathscr{H}(h; \{n_a\}_{a=1}^{\ell-r})$ which are $O(3)$-symmetric and which also satisfy equation (1.14a) has not been catalogued. The most general form of $O(3)$-symmetric (A,Φ) is known, however [24]: Given the homomorphism L, $\mathfrak{g}$ decomposes into a direct sum of irreducible representations of $\mathfrak{su}(2)$. We denote the integer representations $(\Theta_1, \ldots, \Theta_t)$; each has dimension $(2\ell_j + 1)_{j=1}^t$. Each Θ_j is composed of matrices

$$\mathbf{Y}_m^{\ell_j}, \quad m = -\ell_j, -\ell_j + 1, \ldots, \ell_j \quad .$$

With $[L_3, \mathbf{Y}_m^{\ell_j}] = -im\, \mathbf{Y}_m^{\ell_j}$,

$$\sum_{a=1}^{3} [L_a[L_a, \mathbf{Y}_m^{\ell_j}]] = -\ell_j(\ell_j+1)\mathbf{Y}_m^{\ell_j} \quad . \tag{18.8}$$

Denote by $(\{Y_m^{\ell}(\hat{x})\}_{m=-\ell}^{\ell})_{\ell=0}^{\infty}$ the standard spherical harmonics, where $\hat{x}^j = x^j/|x|$.

The most general solution [24] to (18.6) is

$$\Phi(r,\theta,\chi) = \sum_{\alpha=1}^{t} \phi_j(r) \sum_{m=-\ell_\alpha}^{\ell_\alpha} Y_{-m}^{\ell_\alpha}(\phi,\chi)\, \mathbf{Y}_m^{\ell_\alpha} \quad . \tag{18.9}$$

Using the residual gauge freedom in the ansatz (18.6-7), the most general solution to (18.7) may be written down in the following way: Define for $\alpha = 1,\dots,t$,

$$\mathbb{P}_\alpha(\theta,\chi) = \sum_{m=-\ell_\alpha}^{\ell_\alpha} Y_{-m}^{\ell_\alpha}(\theta,\chi)\, \mathbf{Y}_m^{\ell_\alpha} \quad . \tag{18.10}$$

The generic form of the $O(3)$ – symmetric connection as defined by L is:

$$A = \frac{1}{r}\left(\sum_{\alpha=1}^{t}\left(a_{1\alpha}\varepsilon^{inm}[L_m,\mathbb{P}_\alpha]\,\hat{x}^n + a_{2\alpha}[L_i,\mathbb{P}_\alpha]\right) - L_i\right)\varepsilon^{ijk}x^j dx^k \tag{18.11}$$

where $(a_{1\alpha}, a_{2\alpha})_{\alpha=1}^{t}$ are functions of r only.

Define a set $\mathscr{C}(L,h,\{n_a\}_{a=1}^{\ell-r})$ by

$$\mathscr{C}(L,h,\{n_a\}_{a=1}^{\ell-r}) = \Big\{ (A,\Phi) \in \mathscr{H}(h,\{n_a\}_{a=1}^{\ell-r})$$

such that

(1) (A,Φ) is defined by (18.9-11) and the homomorphism L.

(2) (A,Φ) satisfies (1.14a) with $n_a \leq 0$, $(a = 1,\ldots,\ell-r)$ and

$$\sum_{a=1}^{\ell-r} n_a < 0 \quad . \tag{18.12}$$

(3) There exist constants $0 \leq C_0(A,\Phi) < \infty$ and $m(A,\Phi) > 0$ such that

$$\left.\sum_{\alpha=1}^{t} (|a_{1\alpha}|^2 + |a_{2\alpha}|^2)^{1/2} \leq C_0(A,\Phi)\exp(-m(A,\Phi)|x|)\right\} \quad .$$

We remark that if $L: \mathfrak{su}(2) \to \mathfrak{g}$ and h define a set $\mathscr{C}(L,h,\{n_a\}_{a=1}^{\ell-r})$ then necessarily $[L_3,h] = 0$. Further, since $(A,\Phi) \in \mathscr{C}$ is smooth, $\Phi(0) = 0$.

The content of Theorem 18.2 is that given certain sets

$$(A_i,\Phi_i) \in \mathscr{C}(L^{(i)},h,\{n_a^{(i)}\}_{a=1}^{\ell-r})_{i=1}^{N}$$

there exists a solution (A,Φ) to (1.14a) which may be interpreted as a configuration of widely spaced, noninteracting, spherically symmetric monopoles given by $(A_i,\Phi_i)_{i=1}^{N}$.

THEOREM 18.2. *Given a finite set* $\{(A_i,\Phi_i) \in \mathscr{C}(L^{(i)},h,\{n_a^{(i)}\}_{a=1}^{\ell-r})$ *such that* $[L_3^{(i)},L_3^{(j)}] = 0$, $i,j = 1,\ldots,N$, *there exist constants* $\infty > R_0$, d_0, $\varepsilon > 0$ *such that the following is true: For any set of* N *points* $\{x_1,\ldots,x_N\}$ *satisfying*

$$|x_i - x_j| \geq d > d_0 \;, \qquad i \neq j = 1,\ldots,N \;,$$

there exists a solution (A,Φ) *to* (1.14a) *with the properties*

(a) $$(A,\Phi) \in \mathscr{H}\left(h;\left\{\sum_{i=1}^{N} n_a^{(i)}\right\}_{a=1}^{\ell-r}\right) \quad .$$

(b) *In each open ball*

$$B_R(x_i) = \left\{ x \in \mathbb{R}^3 \,\middle|\, |x - x_i| < R_0 \left(\frac{d_0}{d}\right)^{1/2} \equiv R \right\}$$

exists a smooth gauge transformation $g_i : B_R(x_i) \to G$ *such that*

$$|A(x) - g_i A_i(x-x_i) g_i^{-1} - g_i dg_i^{-1}|^2$$
$$+ |\Phi(x) - g_i \Phi_i(x-x_i) g_i^{-1}|^2 < \varepsilon \frac{d_0}{d} \quad . \tag{18.13}$$

If the little group J is the maximal torus, the dimension of the space of moduli to a solution $(A,\Phi) \in \mathcal{H}(h,\{n_a\}_{a=1}^{\ell-r})$ of (1.14a) was calculated by Weinberg [23]. He found that if ℓ' of the integers $n_a = 0$ and $\ell - \ell'$ of the $n_a > 0$, the dimensions of the space of moduli is $4\,\Sigma_a n_a - \ell + \ell'$. This is partially verified by the following corollary to Theorem 18.2.

COROLLARY 18.3. *There exists at least a* $3\,\Sigma_{a=1}^{\ell-r} n_a$ *dimensional sublattice of distinct, gauge inequivalent solutions to* (1.14a) *in* $\mathcal{H}(h,\{n_a\}_{a=1}^{\ell-r})$.

We refer the reader to [19] for the proofs of Theorems 18.1,2 and Corollary 18.3. The basis of the proofs is Theorems 2.1 and 2.2.

We state without proof the following theorem concerning the asymptotic decay of the field strengths.

THEOREM 18.4. *Let* (A,Φ) *be a smooth, finite action solution to* (1.2-3) *with* G *any compact Lie group and* $\lambda \geq 0$. *Then*

$$|F_A| + |D_A\Phi| \leq \text{const}(1 + |x|)^{-2} \quad , \tag{18.14}$$

and

$$0 < 1 - |\Phi| \leq \text{const}(1 + |x|)^{-1} \quad . \tag{18.15}$$

An immediate corollary following from Theorem 18.4 and Theorem II.3.1 is

COROLLARY 18.5. *Let* (A,Φ) *be a smooth, finite action solution to* (1.4) *with* G *any compact Lie group and* $\lambda \geq 0$. *Then* $(A,\Phi) \in \mathscr{H}(h; \{n_a\}_{a=1}^{\ell-r})$ *for some unit vector* $h \in g$ *and integers* $\{n_a\}_{a=1}^{\ell-r}$.

V. SMOOTHNESS

V.1. THE SMOOTHNESS PROBLEM

Sufficient conditions to conclude that a solution to the Yang-Mills-Higgs equations on $\mathbb{R}^d$, $d = 2,3$ is gauge equivalent to a smooth solution are established. The conditions are stronger than the assumption of finite action. We assume that there exists a gauge in which the solution is locally square integrable, with locally square integrable first derivatives. The smoothness theorem follows K. Uhlenbeck's existence proof of a local Coulomb gauge; in this gauge the Yang-Mills-Higgs equations are uniformly elliptic, a situation in which much is known.

The main result of this chapter is

THEOREM 1.1. *Take* G *to be a compact, connected Lie group. Let* $d = 2$ *or* 3. *Let* (A,Φ) *be a finite action, distribution solution to the Yang-Mills-Higgs equations* (2.3) *below. Suppose that the components of* A *and* Φ *are locally integrable with locally square integrable first derivatives. Then there is a Hölder continuous gauge transformation which makes* (A,Φ) *smooth on* $\mathbb{R}^d$. *Further,* (A,Φ) *is locally gauge equivalent to a real analytic configuration.*

Remark. Theorem 1.1 is proved for $d = 2,3$. It is a consequence of recent work by Uhlenbeck [6] that the theorem is true for $d = 4$ with "$L^2_{2(Loc)}(\mathbb{R}^4; G)$" substituted for "Hölder continuous."

The theorem is also true if $\mathbb{R}^d$ is replaced by a compact Riemannian manifold with real analytic structure. If the manifold has no real analytic structure, then the theorem applies, except for the last sentence. (See Theorem 2.4.)

The compact case is treated first, in Sections 2-5. The extension to Euclidean spaces is made in Section 6.

Some useful *a priori* estimates of a global nature are derived in §7 and §8.

There are unsolved questions concerning the nature of singular solutions:

CONJECTURE 1.2. *The conclusions of Theorem* 1.1 *hold under the weaker assumption that* (A,Φ) *is a finite action solution to* (2.2).

Problem 1.3. Without the assumption of finite action, characterize the possible singularities a distribution solution may have.

V.2. THE EQUATIONS ON COMPACT MANIFOLDS

Before proving Theorem 1.1, we consider the Yang-Mills-Higgs equations on compact domains where the smoothness problem is essentially a local one. We state the main results in this section. Let M be a compact, connected and oriented Riemannian manifold of dimension 2 or 3. Let G be a compact, connected Lie group and let $p: P \to M$ be a principle G-bundle over M. (In this notation, p is the canonical projection onto M.) Let $p: E \to M$ be a vector bundle associated to P by a smooth representation ρ of G. Recall that the Higgs field is a section of E and the covariant derivative of Φ, denoted $D_A\Phi$, is a section of $T^* \otimes E$. (See Chapter VI.5.) Meanwhile the curvature F_A of a connection A on P is a section of the bundle $\Lambda_2 T^* \otimes \mathrm{ad}(P)$. Here $\mathrm{ad}(P)$ is the vector bundle associated to P via the adjoint representation.

As in Chapters II.1 and VI.5, let $(\ ,\ )$ denote an inner product on E with respect to which the representation ρ is unitary. We also use $(\ ,\ )$ to denote the inner products on the vector bundles $\Lambda_k T^* \otimes \mathrm{ad}(P)$, $0 \le k \le d$. The inner products on $\Lambda_k T^*$ are induced by the Riemannian metric on M in the usual way.

If $U \subseteq M$ is an open set, then the action over U is

$$\mathscr{A}_U(A,\Phi) = \frac{1}{2} \int_U \left\{ (F_A,F_A) + (D_A\Phi,D_A\Phi) + I(|\Phi|^2) \right\} \omega \tag{2.1}$$

Here ω is the volume form on M. We take as the function $I(t)$ a positive, real-valued polynomial of degree ν.

DEFINITION 2.1. *A connection* A *and Higgs field* Φ *is a* *distribution solution to the Yang-Mills-Higgs equations over* $U \subseteq M$ *iff for all* $(b,\eta) \in C_0^\infty(U;T^* \otimes \text{ad } P) \oplus C_0^\infty(U;E)$,

$$0 = \text{grad } \mathscr{A}[A,\Phi;b,\eta] \equiv \int_U \Big\{ (D_A b,F_A) + (\rho(b)\Phi,D_A\Phi)$$

$$+ (D_A\eta,D_A\Phi) + I'(|\Phi|^2)(\eta,\Phi) \Big\} \omega \quad . \tag{2.2}$$

Here $\rho(\cdot)$ *is the induced representation of* G *on* $T^*\otimes \text{ad } P$. *The form* $\rho(b)$ *is a section of* $T^*\otimes \text{End}(E)$; *a one-form with values in the vector space of endomorphism of* E. *The function* $I'(t) = \frac{d}{dt} I(T)$ *is a polynomial of degree* $\nu-1$.

DEFINITION 2.2. *A connection* A *is an* L_p^k *connection over* $U \subseteq M$ *iff for each* $x \in U$ *there is a neighborhood* $B(x) \subset U$ *of* x *and a smooth section* $g\colon B(x) \to P$ *such that the Lie algebra valued* 1*-form* $g^*A \in L_p^k(U;T^* \otimes \mathfrak{g})$. *Equivalently: Let* $\{x_j\}_{j=1}^d$ *be local coordinates in* $B(x)$ *and* $\{h^a\}_{a=1}^{\ell}$, $\ell = \dim \mathfrak{g}$, *be a basis of the Lie algebra* $\mathfrak{g}$. *Then the components of* $g^*A = A_j^a h^a dx^j$ *satisfy*

$$\|A\|_{k,P;B(x)} \equiv \left[\int_{B(x)} \sum_{0\le|s|\le k} (\nabla^{(s)} A_j^a \nabla^{(s)} A_j^a)^{p/2} \, \omega \right]^{2/p} < \infty \, . \tag{2.3}$$

DEFINITION 2.3. *A section* Φ *of* E *is an element of* $L_p^k(U;E)$ *iff for each* $x \in U$ *there is a neighborhood of* x, $B(x) \subset U$ *and a smooth orthonormal basis* $\{e_a\}_{a=1}^n$, $n =$ *(dimension of fiber of* E*), over* $B(x)$

such that the components $\Phi^a = (e_a, \Phi)$ *are in* $L_p^k(U)$. *That is,*

$$\|\Phi\|_{k,P;B(x)} \equiv \left[\int_{B(x)} \sum_{0\le|s|\le k} (\nabla^{(s)}\Phi^a \nabla^{(s)}\Phi^a)^{p/2}\,\omega\right]^{2/p} < \infty . \quad (2.4)$$

Remark. The notion of a connection or Higgs field being in L_p^k is invariant under smooth gauge transformations. In fact, for $p \ge 2$, the statement $(A,\Phi) \in L_p^k$ is invariant under gauge transformations with components locally in L_p^{k+1}.

Remark. In the cases where $M = \mathbb{R}^d$, $d = 2,3$, these definitions simplify. The statement A is a L_p^k connection is just the statement that the components of A in the decomposition $A = A_j^a dx^j h^a$ are in the Sobolev space L_p^k (see Chapter VI.1). Similarly $\Phi \in L_p^k$ means that the components Φ^a with respect to an orthonormal basis are in L_p^k.

In the case where M and P are not topologically trivial, there are no global orthonormal sections. Hence one must define the differential properties of connections and cross-sections in terms of locally orthonormal frames. This is all that Definitions 2.2-3 achieve.

THEOREM 2.4. *Let* (A,Φ) *be respectively an* $H_1 = L_2^1$ *connection on* P *and an* H_1 *section of* E, *over an open set* $U \subseteq M$. *Suppose that in addition,* $\mathscr{A}_U(A,\Phi) < \infty$ *and* (A,Φ) *is a distribution solution to the Yang-Mills-Higgs equations over* U. *Then* (A,Φ) *is gauge equivalent over every open set* $V \subseteq U$ *disjoint from* $\partial\bar{U}$ *to* $(A_g, \Phi_g) = (gA_g^{-1} + gdg^{-1}, g\Phi g^{-1})$ *which is smooth. Further,* A_g *is a connection on* P, Φ_g *is a section of* E *and* g *is Hölder continuous. The configuration* (A_g, Φ_g) *is a strong solution to the Yang-Mills-Higgs equations.*

COROLLARY 2.5. *Let* (A,Φ) *be as in Theorem* 2.4. *Suppose that* P *and* M *are real analytic. About each* $x \in U$ *there is a neighborhood of* x, $B(x) \subseteq U$ *and a Hölder continuous gauge transformation* h *on* $B(x)$ *such that* (A_h, Φ_h) *are real analytic in* $B(x)$.

Remark. With some changes, the above Theorem and Corollary are true in 4-dimensions. (See Uhlenbeck.) In $d = 4$, the gauge transformation g

is not necessarily continuous. Thus while (A_g, Φ_g) are smooth, one cannot conclude that A_g is a smooth connection on the *same* principal G-bundle P on which the problem was originally formulated.

V.3. LOCAL REAL ANALYTICITY

In this section we prove that the open set $V \subseteq U \subseteq M$ which is disjoint from ∂U is covered uniformly by a finite number of balls of fixed radius such that in *each* ball Theorem 2.4 holds.

PROPOSITION 3.1. *Let* (A,Φ) *satisfy the conditions of Theorem 2.4. Let* $V \subseteq U \subseteq M$ *be as stated. Every point* $x \in V$ *is contained in some open ball* $B(x)$ *in which* (A,Φ) *is gauge equivalent to a smooth configuration. If* P *is real analytic then in* B_x, (A,Φ) *is gauge equivalent to a real analytic configuration.*

Proof. We use the fact that $\mathrm{grad}\, \mathscr{A}[A,\Phi;b,\eta] = 0$.

LEMMA 3.2. *Let* (A,Φ) *be a finite action,* L_2^1 *connection and Higgs field in a ball* $B \subset M$. *Assume that* (A,Φ) *is a distribution solution to the Yang-Mills-Higgs equations over* B. *Then* $\mathrm{grad}\, \mathscr{A}[A,\Phi;\cdot]$ *extends to a bounded, linear functional on the Banach space*

$$H \equiv H_{1(0)}(B;T^* \otimes \mathrm{ad}P) \oplus (H_{1(0)}(B;E) \cap L_{2\nu}(B;E)) \quad .$$

Further, $\mathrm{grad}\, \mathscr{A}[A,\Phi;\cdot] \equiv 0$ *on* H.

Proof. For $(b,\eta) \in C_0^\infty$, Hölder's inequality gives

$$|\mathrm{grad} \quad [A,\Phi;b,\eta]| \le \|F\|_{L_2} \|D_A b\|_{L_2} + \mathrm{const}\|b\|_{L_4} \|\Phi\|_{L_4} \|D_A\Phi\|_{L_2}$$

$$+ \|D_A\eta\|_{L_2} \|D_A\Phi\|_{L_2} + \|I'(|\Phi|^2)(\Phi,\eta)\|_{L_1} \quad . \tag{3.1}$$

All the norms above are over B. Using the triangle inequality, we obtain

$$\|D_A b\|_{L_2} \leq \|b\|_{H_1} + \|Ab\|_{L_2} \leq \|b\|_{H_1} + \|A\|_{L_4} \|b\|_{L_4}$$

$$\leq \text{const}\cdot \|b\|_{H_1}\left(1 + \|A\|_{H_1}\right) \quad . \tag{3.2}$$

The last line is a Sobolev inequality. Similarly,

$$\|D_A \eta\|_{L_2} \leq \text{const}\|\eta\|_{H_1}\left(1 + \|A\|_{H_1}\right) \quad . \tag{3.3}$$

As for the term with I', we remark that if $0 \leq I(t)$ is a polynomial of degree ν, there exist constants $0 < c_1, c_2 < \infty$ such that

$$\frac{1}{2} c_1 I(t) \geq t^{\nu} - c_2 \quad . \tag{3.4}$$

Since the action is finite, $\Phi \in L_\nu$ and

$$\|\Phi\|_{L_{2\nu}}^{2\nu} \leq c_1 \mathscr{A}(A,\Phi) + c_2 \quad . \tag{3.5}$$

Using Hölder's inequality,

$$\|I'(\Phi)(\Phi,\eta)\|_{L_1} \leq \|I'(\Phi)\Phi\|_{L_{2\nu/(2\nu-1)}} \|\eta\|_{L_{2\nu}} ,$$

$$\text{const}\left(1 + \|\Phi\|_{L_{2\nu}}^{2\nu}\right)\|\eta\|_{L_\nu} \quad . \tag{3.6}$$

Therefore, using (3.2-6) in (3.1) we obtain

$$|\text{grad}\,\mathscr{A}[A,\Phi;b,\eta]| \leq \text{const}\left(\|b\|_{H_1} + \|\eta\|_{H_1} + \|\eta\|_{L_{2\nu}}\right), \tag{3.7}$$

where the constant depends on (A,Φ). The functional now extends by continuity.

PROPOSITION 3.3. (Uhlenbeck [6].) *Let* $B_r \subset U$ *be an open ball of radius* $r > 0$. *Suppose that* A *is an* L_2^1 *connection over* B_r. *There exist constants* $0 < \beta$, $k < \infty$ *such that if* $\|A\|_{H_1;B_r} < \beta$, *then there exists a gauge* $g\colon B_r \to G$ *with* H_2 *components such that* $A_g = gAg^{-1} + gdg^{-1}$ *satisfies*

(1) $\|A_g\|_{H_1;B_r} < k\beta$.

(2) $d{*}A_g = 0$.

(3) $i^*({*}A_g) = 0$. (3.8)

Here $i\colon \partial\bar{B}_r \to B_r$ is the inclusion and i^* denotes the pull back map induced by i.

Remark. For the dimensions of M considered, the gauge transformation g of the proposition is Hölder continuous (Corollary VI.2.7b).

From Proposition 3.2, V can be covered uniformly by a finite number of balls $\{B_r(i) \subseteq U\}_{i=1}^{\ell}$ such that there exists $g_i \in H_2$, for each i, with $A(i) = A_{g_i}$ satisfying 1, 2 and 3 of that proposition. We assume that the set $\{B_{r/2}(i)\}_{i=1}^{\ell}$ covers V also.

Fix attention on one such ball $B_r(1) = B_r$. Let $A = A(1)$ and $\Phi = \Phi(1) = \rho(g_1)\Phi$.

Fix local coordinates $\{x^j\}_{j=1}^{d}$ in B_r and expand all forms with respect to the basis $\{dx^j\}_{j=1}^{d}$ of $C^\infty(B_r;T^*)$. Denote by $\alpha^{ij}(x)$ the Riemannian metric in B_r, R_{ij} the Ricci tensor and ω_M the volume form.

Define $v \in H_1(B_r) \cap L_\nu(B_r)$ by

$$v = (1 + (\Phi,\Phi))^{1/2} , \qquad (3.9)$$

and define $w \in H_1(B_r)$ by

$$w = (1 + \alpha^{ik}(A_j,A_k))^{1/2} , \qquad (3.10)$$

where in B_r, $A = A_j dx^j$ with A_j Lie algebra valued. We derive differential inequalities for v and w. Note that because I(t) is a positive polynomial of finite degree, $I_0 = \min_{t\geq 0} I'(t)$ is defined.

LEMMA 3.4. *Let* w *and* v *be as above. For all* $0 \leq \psi \in H_{1(0)}(B_r)$,

$$\int_{B_r} \omega\{\alpha^{ij}\nabla_i\psi\nabla_j v + \psi v I_0\} \leq 0 \quad . \tag{3.11}$$

$$\int_{B_r} \omega\{\alpha^{ij}\nabla_i\psi\nabla_j w + (\psi\nabla_i w - w\nabla_i\psi)e^i + \psi w f\} \leq 0 \quad , \tag{3.12}$$

where

$$e^i = \alpha^{ij}\omega^{-2}(A_k, A_\ell)\nabla_j\alpha^{k\ell} \quad ,$$

$$f = \omega^{-2}\alpha^{k\ell}\alpha^{ij}\{(A_k,[F_{A_{j\ell}},A_i]) + (\nabla_i A_k,[A_j,A_\ell]) + (A_i,R_{jk}A_\ell)\}$$

$$+ \omega^{-2}\alpha^{ij}(\rho(A_i),\nabla_{A_j}\Phi) \quad . \tag{3.13}$$

Proof. Let $0 \leq \psi \in H_{1(0)}(B_r) \cap L_\nu$ be arbitrary. Define a section $\eta \in H_{1(0)} \cap L_\nu$ by

$$\eta = \psi\Phi v^{-1} \quad . \tag{3.14}$$

Then,

$$dv = v^{-1}(\Phi,\nabla_A\Phi) \quad . \tag{3.15}$$

The equation resulting from the condition grad $\mathscr{A} = 0$ is

$$\int_{B_r} \omega\left\{(\nabla\psi,\nabla v) + \psi v^{-1}(|\nabla_A\Phi|^2 - |\nabla v|^2) + \psi v I'(|\Phi|^2)|\Phi|^2 v^{-2}\right\} = 0 \quad . \tag{3.16}$$

Equation (3.11) follows upon noticing that

$$|\nabla_A\Phi|^2 - |\nabla v|^2 \geq 0 \quad . \tag{3.17}$$

In order to prove (3.12), define $b \in H_{1(0)}$ by

$$b = b_i dx^i \quad , \qquad \text{where} \quad b_i = \psi A_i w^{-1} \quad . \tag{3.18}$$

Then because

$$w\nabla_j w = (A_i, \nabla_j A_k)\alpha^{ik} + (A_i, A_k)\nabla_j \alpha^{ik} \quad ,$$

we obtain

$$\nabla_j b_k = (\nabla_j \psi) A_k w^{-1} + \psi(\nabla_j A_b) w^{-1} - \psi A_k (\nabla_j w) w^{-2} \ . \tag{3.19}$$

The above identities allow the computation of the equation $\operatorname{grad}\mathscr{A}[A,\Phi;b,0] = 0$. It is here that the gauge condition $d{*}A = 0$ is used. In fact, for $b \in H_{1(0)}$,

$$0 = \int_{B_r} \omega \left\{ (D_A b, F_A) + (\rho(b)\Phi, \nabla_A \Phi) + (D_A {*}b, D_A {*}A) \right\} \quad , \tag{3.20}$$

as well as (2.2). In terms of the components A_i and b_i, equation (3.20) reads

$$\int_{B_r} \omega_M \Big\{ \alpha^{ij}\alpha^{k\ell} [(\nabla_i b_k, \nabla_j A_\ell) + (b_k, [F_{Ak\ell}, A_i]) + (\nabla_i b_k, [A_i, A_\ell]) + (b_i, R_{jk} A_\ell)] + \alpha^{ij}(\rho(b_i)\Phi, \nabla_{A_j}\Phi) \Big\} = 0 \ . \tag{3.21}$$

Substituting of (3.19) into (3.21) yields (3.12). This is a straightforward calculation which we omit.

LEMMA 3.5. *Both* v *and* w *are bounded in* $B_{r-\varepsilon}$ *for any* $\varepsilon > 0$.

Proof. For v we appeal to Theorem VI.9.2 directly, since (3.11) is satisfied. As for w, we use (3.12). Note that $e^i \in L_\infty$ and $f \in L_2$. Thus, for any $B_\delta \subset B_r$,

$$\int_{B_\delta} (|e| + |f|)^{d/2} \le \text{const } \delta^\mu \quad , \tag{3.22}$$

where $\mu = (4-d)/4d$. Thus by Theorem VI.9.2, $w \in L_\infty(B_{r-\varepsilon})$ also.

In the next step of the proof, each component of (A,Φ) is considered individually. Let $\{h^\mu\}_{\mu=1}^{\dim \mathfrak{g}}$ be a basis for $\mathfrak{g}$ and $\{\ell^a\}_{a=1}^n$ be a basis for $\mathbb{L}$. Denote the structure of G by

$$g^{\mu\nu\sigma} = (h^\mu, [h^\nu, h^\sigma]) \tag{3.23a}$$

and the representation matrix of ρ by

$$\rho^{a\mu b} = (\ell^a, \rho(h^\mu)(\ell^b)) \quad . \tag{3.23b}$$

Expand $A_j = A_j^\mu h^\mu$ and $\Phi = \Phi^a \ell^a$ in terms of the basis $\{h^\mu\}$ and $\{\ell^a\}$, respectively.

LEMMA 3.6. *For all* $\psi \in H_{1(0)}(B_r)$,

$$\int_{B_r} \omega \left\{ \alpha^{ij} \Delta_i \psi \nabla_j \Phi^1 + \nabla_i \psi e^i + \psi f \right\} = 0 \quad ,$$

with $e^i = \alpha^{ij} \rho^{1\mu a} A_j^\mu \Phi^a$,

$$f = \alpha^{ij} (\rho^{a\mu 1} A_i^\mu \nabla_j \Phi^a + \rho^{a\nu 1} \rho^{a\mu b} \Phi^b A_i^\nu A_j^\mu) + I' \Phi^1 \tag{3.24}$$

<u>Proof</u>. Expand the equation $0 = \operatorname{grad} \mathscr{A}[A,\Phi;0,\eta]$ with $\eta = \psi \ell^1$.

LEMMA 3.7. *For all* $\psi \in B_r$,

$$\int_{B_r} \omega_M \{ \alpha^{ij} \nabla_i \psi \nabla_j A_1^1 + \nabla_i \psi e^i + \psi f \} = 0 \quad ,$$

with

$$e^i = \alpha^{ij} g^{1\mu\nu} A_j^\mu A_1^\nu \quad ,$$

and

$$f = \alpha^{ij}\Big(\nabla_i \alpha_{1k}\alpha^{k\ell}\nabla_j A^1_\ell + g^{1\mu\nu} F_{A_{j1}} A^\mu_i$$

$$+ \alpha^{k\ell}\nabla_i \alpha_{1k} g^{1\mu\nu} A^\mu_j A^\nu_\ell + R_{1i} A^1_j\Big) + \rho^{alb}\Phi^a \nabla_{A_1}\Phi^b \quad . \qquad (3.25)$$

Here $\alpha_{\ell k} = (\alpha^{-1})^{\ell k}$.

Proof. Expand (3.20) using $b = \psi dx^1 \sigma^1$.

LEMMA 3.8. *The components* Φ^1 *and* A^1_1 *are in* $H_2(B_{r-\varepsilon})$ *for any* $\varepsilon > 0$.

Proof. We appeal to Theorem VI.9.1. Consider first the equation for Φ^1. The conditions of Theorem VI.9.1 are satisfied when $e^i \in L_\infty(B_{r-\varepsilon}) \cap H_1(B_{r-\varepsilon})$ and $f \in L_2(B_{r-\varepsilon})$ with e^i, f defined in (3.25). But, $e^i \in L_\infty \cap H_1$ from Lemma 3.8. The same Lemma implies that $\|f\|_{L_2;B_{r-\varepsilon}} < \infty$ also. We conclude $\Phi^1 \in H_2(B_{r-\varepsilon})$.

As for A^1_1, the functions e^i, f defined in (3.25) satisfy

(1) $E_i \in L_\infty(B_{r-\varepsilon}) \cap H_1(B_{r-\varepsilon})$ and

(2) $f \in L_2(B_{r-\varepsilon})$.

Both statements follow from Lemma 3.8 also. Theorem VI.9.1 gives the lemma.

Similar analysis of the equations for the components Φ^a, $a = 2,\dots,n$ and A^μ_j, $\mu = 2,\dots,\dim \mathcal{G}$, $j = 1,\dots,d$ leads to the conclusion that $A, \Phi \in H_2(B_{r-\varepsilon})$, $\varepsilon > 0$. We omit the details.

At this point, the iteration procedure to prove that $(A,\Phi) \in H_k(B_{r-\varepsilon})$, $k > 0$ is standard. See Morrey, Chapter 6. We give a version. The proof is by induction on k.

PROPOSITION 3.9. Suppose that for every $\varepsilon > 0$, $(A,\Phi) \in H_k(B_{r-\varepsilon})$, $k \geq 2$, further, suppose that A satisfies (3.8) and that (A,Φ) is a distribution solution of the Yang-Mills-Higgs equations on B_r. Then $(A,\Phi) \in H_{k+1}(B_{r-\varepsilon})$ for every $\varepsilon > 0$.

Remark. Proposition 3.1 follows since $H_k \to C^{k-2}$ by a Sobolev inequality, Corollary VI.2.7. Real analyticity follows if P is real analytic by standard elliptic regularity, see Morrey.

Proof. We consider (3.24) for $\Phi^1 \in H_k$, $k \geq 2$. Then for any $\varepsilon > 0$.

$$-\Delta_M \Phi^1 - \nabla_i e^i + f = 0, \quad (\text{a.e. in } B_{r-\varepsilon}) \quad . \tag{3.26}$$

Here Δ_M is the Laplace-operator on M defined by the metric. Both e^i, f are given by (3.24). For $k > d/2$, H_k is a Banach algebra, this is to say that if $a, b \in H_k$ then $ab \in H_k$. Therefore, $f, \nabla_i e^i \in H_{k-1}$. Let $s = (s_1, \ldots, s_d)$ be a multi-index with $|s| = k-1$. Then for all $\psi \in H_{1(0)}(B_{r-\varepsilon})$,

$$0 = \int_{B_{r-\varepsilon}} \omega \left\{ \alpha^{ij} \nabla_i \psi \nabla_j \nabla^{(s)} \Phi^1 + \psi(\nabla^{(s)}(-\nabla_i e^i + f) + [\nabla^{(s)}, \Delta_M]\Phi^1) \right\} \tag{3.27}$$

Note that $\nabla^{(s)}\Phi^1 \in H_1(B_{r-\varepsilon/2})$ and $\nabla^{(s)}(\nabla_i e^i - f) \in L_2(B_{r-\varepsilon})$ by assumption. Further, since the metric is smooth, $[\nabla^{(s)}, \Delta_M]\Phi^1 \in L_2(B_{r-\varepsilon})$ also, since it involves derivatives of the metric components α^{jk} and derivatives of Φ^1 of order k and smaller. Thus by Theorem VI.9.1, $\nabla^{(s)}\Phi \in H_2(B_{r-\varepsilon})$, that is, $\Phi^1 \in H_{k+1}(B_{r-\varepsilon})$. Similar arguments hold for Φ^a, $a = 2, \ldots, N$ and A_j^μ, $j = 1, \ldots, d$; $\mu = 1, \ldots, \dim \mathfrak{g}$.

V.4. PRINCIPAL AND ASSOCIATED BUNDLES

This section is a review of bundle theory [1], following Mitter. If $p: P \to M$ a principal G-bundle over M, then P comes equipped with a bundle atlas, $\{U_\alpha, h_\alpha\}_{\alpha \in \Lambda}$, ω here $\{U_\alpha\}$ is an open cover of M and

$$h_\alpha : p^{-1}(U_\alpha) \to U_\alpha \times G \quad . \tag{4.1a}$$

The maps h_α satisfy

$$h_\alpha(yg) = h_\alpha(y)g, \qquad y \in P, \quad g \in G. \tag{4.1b}$$

Thus, the h_α give local coordinates for P over U_α. If $U_\alpha \cap U_\beta = \emptyset$ then $h_\alpha \circ h_\beta^{-1}$ define the transition functions,

$$h_{\alpha\beta}: U_\alpha \cap U_\beta \to G \quad ,$$

by

$$(x, h_{\alpha\beta}(x)) = h_\alpha \circ h_\beta^{-1}((x,e)) \quad . \tag{4.2}$$

Here e is the identity element of G. It follows that if $y \in P$, $p(y) \in U_\alpha \cap U_\beta$ and $h_\beta(y) = (x,g)$ then $h_\alpha(y) = (x, h_{\alpha\beta}g)$.

It is a fundamental theorem on the subject that a principal G bundle is uniquely determined by its set of transition functions and vice-versa. By definition, a bundle is C^k iff the transition functions are C^k. See Kobayashi and Nomizu, Proposition I.5.2.

A C^k global gauge transformation g is equivalent to a set of C^k maps $\{g_\alpha: U_\alpha \to G\}$ which satisfy where $U_\alpha \cap U_\beta \neq \emptyset$,

$$g_\alpha h_{\alpha\beta} = h_{\alpha\beta} g_\beta \quad . \tag{4.3}$$

In fact, where $U_\alpha \cap U_\beta \neq \emptyset$,

$$h_\beta^{-1}(x, g_\beta h_{\alpha\beta}^{-1} g) = h_\alpha^{-1}(x, g_\alpha g) \quad . \tag{4.4}$$

Hence $\{g_\alpha\}$ defines a global automorphism of P which commutes with the right action of G and the projection to M.

An associated vector bundle E is defined as follows. Let $\mathbb{L}$ be a vector space on which a smooth representation ρ of G is defined. Then E is the quotient of $P \times \mathbb{L}$ under the identification

$$(yg, \rho(g^{-1})\ell) \sim (y, \ell) \quad . \tag{4.5}$$

The transition functions for E are the set of maps $\rho(h_{\alpha\beta})\colon U_\alpha \cap U_\beta \to \mathrm{Au} + \mathbb{L}$, hence E is C^k if P is.

A section Φ of E is equivalent to a set $\{\Phi_\alpha\colon U_\alpha \to \mathbb{L}\}$ subject to the requirement that where $U_\alpha \cap U_\beta \neq \emptyset$,

$$\Phi_\alpha = \rho(h_{\alpha\beta})(\Phi_\beta) \quad . \tag{4.6}$$

A C^k connection A, on P is equivalent to a set of Lie algebra valued 1-forms $\{A_\alpha \in C^k(U_\alpha; T^* \otimes \mathfrak{g})\}$, such that where $U_\alpha \cap U_\beta \neq \emptyset$,

$$A_\alpha = h_{\alpha\beta} A_\beta h_{\alpha\beta}^{-1} + h_{\alpha\beta} dh_{\alpha\beta}^{-1} \quad . \tag{4.7}$$

Equation (4.7) must be true in order that the covariant derivative ∇_{A_j} map sections of E into sections of E.

Under the action of a gauge transformation $g = \{g_\alpha\}$, a section Φ of E transforms as

$$\Phi_\alpha \to \rho(g_\alpha)\Phi_\alpha \quad , \tag{4.8}$$

and a connection as

$$A_\alpha \to g_\alpha A_\alpha g_\alpha^{-1} + g_\alpha dg_\alpha^{-1} \quad . \tag{4.9}$$

We introduce the notion of an isomorphism of G-bundles:

DEFINITION 4.1. *Two* C^k *principal* G *bundles* $p\colon P \to M$, $p'\colon P' \to M$ *are* C^ℓ *isomorphic* $(\ell \leq k)$ *if there exists a* C^ℓ *homeomorphism*

$$\theta\colon P \to P' \quad ,$$

with the properties

(1) θ *commutes with right multiplication,* $\theta(yg) = \theta(y)g$ *for* $y \in P$, $g \in G$.

(2) θ *commutes with the projections,* $\theta(py) = p'\theta(y)$.

DEFINITION 4.2. *If* E *is a vector bundle which is associated to* P *by a representation* ρ *of* G, *then* θ *induces an isomorphism between* E *and* E_θ, *where* E_θ *is the vector bundle associated to* P' *by the same representation* ρ. *The isomorphism* $\theta\colon P\times\mathbb{L}\to\theta(R)\times\mathbb{L}-P'\times\mathbb{L}$ *passes to the quotient,* (4.5) *due to properties* (1) *and* (2) *above. This serves to define* E_θ.

Properties (1) and (2) of Definition 4.1 imply that θ is equivalent to a set of C^ℓ maps $\{\theta_\alpha\colon U_\alpha\to G\}$ such that

$$\theta_\alpha h_{\alpha\beta} = h'_{\alpha\beta}\theta_\beta \quad , \qquad x\in U_\alpha\cap U_\beta . \tag{4.10}$$

Here, $h_{\alpha\beta}$, $h'_{\alpha\beta}$ are the transition functions for P, P'.

If $\Phi=\{\Phi_\alpha\}$ is a section of an associated vector bundle E then $\Phi_\theta=\{\rho(\theta_\alpha)\Phi_\alpha\}$ is a section of $E'=E_\theta$. Meanwhile, if $A=\{A_\alpha\}$ is a connection on E, then $A_\theta=\{\theta_\alpha A_\alpha\theta_\alpha^{-1}+\theta_\alpha d\theta_\alpha^{-1}\}$ is a connection on E_θ.

Using (4.10) an elegant characterization of θ is possible. If P and P' are two principal G-bundles over M then $P\times P'$ is a principal $G\times G$ bundle. Define the associated bundle of groups, Iso(P,P') as the quotient of $P\times P'\times G$ under the equivalence

$$(yg, y'g', g'^{-1}hg) \sim (y, y', h) \quad , \tag{4.11}$$

for $y\in P$, $y'\in P'$, and $h,g,g'\in G$.

PROPOSITION 4.3. *Two bundles* P,P' *over* M *are* C^k *isomorphic iff* Iso(P,P') *admits a global* C^k *section.*

Proof. Suppose Iso(P,P') admits a global C^k section θ. Then θ is equivalent to a set of C^k maps $\{\theta_\alpha\colon U_\alpha\to G\}$ such that where $U_\alpha\cap U_\beta\neq\emptyset$, (4.10) is satisfied. Thus θ is a C^k isomorphism. Similarly a C^k isomorphism by construction, a section of Iso(P,P').

With this terminology, a C^ℓ gauge transformation is a global C section of Iso(P,P). In addition, if $\theta_1,\theta_2\in C^\ell(M, \mathrm{Iso}(P,P'))$, then

$\theta_2^{-1} \circ \theta_1 \in C^{\ell}(M, \mathrm{Iso}(P,P))$ is a global gauge transformation.

In the context of isomorphisms between smooth bundles, a fundamental result is that C^k, $k > 0$ implies C^{∞}.

THEOREM 4.4. *Let* G *be a compact Lie group. Suppose that* P_1 *and* P_2 *are* C^{∞} *principal* G*-bundles over an open set* $V \subseteq M$*, where* M *is a compact, Riemannian manifold. If* P_1 *and* P_2 *are* C^{ℓ} $(\ell > 0)$ *isomorphic then they are* C^{∞} *isomorphic.*

Proof. This result is well known to differential geometers. As per Proposition 4.3, it is enough to show that if $\mathrm{Iso}(P,P')$ admits a C^{ℓ} section over $V \subseteq M$, then it admits a C^{∞} section also.

For the purposes of the proof we consider G as a smooth, compact submanifold of $\mathbb{M}(m,\mathbb{C})$, the vector space of complex $m \times m$ matrices. This is true for m sufficiently large. The group G has a right and left action (matrix multiplication) on $\mathbb{M}(m,\mathbb{C})$ so we construct the associated vector bundle $E_{\mathbb{M}}(P,P')$. This vector bundle is defined to be the quotient of $P \times P' \times \mathbb{M}$ under the equivalence

$$(y, g, y'g', g'^{-1}zg) \sim (y, y', z) \quad , \tag{4.12}$$

for $y \in P$, $y' \in P'$ and $z \in \mathbb{M}$.

The fiber bundle $\mathrm{Iso}(P,P') \subset E_{\mathbb{M}}(P,P')$ is a smooth subbundle with fiber G. Let $g \in C^{\ell}(V; \mathrm{Iso})$ be the given section. Consider g as a section of $E_{\mathbb{M}}$. Since $E_{\mathbb{M}}$ is a C^{∞} vector bundle, there exists $g_{\varepsilon} \in C^{\infty}(V, E_{\mathbb{M}})$ such that

$$|g(x) - g_{\varepsilon}(x)| < \varepsilon \quad . \tag{4.13}$$

(See, e.g., Steenrod, Section 6.7.) Here, $|\cdot|$ is any fixed norm on the fibers, $\mathbb{M}$. Equation (4.13) means that in each U_{α},

$$|g_{\alpha}(x) - g_{\varepsilon\alpha}(x)| < \varepsilon \quad . \tag{4.14}$$

To continue we introduce the notion of a bundle tubular neighborhood. The definition we quote is from Palais, Ch. 12.

DEFINITION 4.5. *Let* E_1 *be a* C^∞ *fiber bundle over* M *and let* E_2 *be a* C^∞*, closed subbundle of* E_1*. By a bundle tubular neighborhood of* E_2 *in* E_1 *we mean a* C^∞ *vector bundle* $r: \mathcal{O} \to E_2$ *over* E_2 *such that* $\mathcal{O}$ *is an open subset of* E_1*, and in fact an open subbundle of* E_1*, and the projection* r *is a* C^∞ *fiber bundle morphism.*

The fundamental existence theorem is

THEOREM 4.6 (Palais). *Let* E_1 *be a paracompact* C^∞ *fiber bundle over* M *and let* E_2 *be a closed* C^∞ *subbundle of* E_1*. Then* E_2 *has a bundle tubular neighborhood in* E_1*.*

Our Theorem 4.6 is Theorem 12.12 of Palais [3]. Letting $E_1 = E_{\mathbb{M}}(P,P')$, $E_2 = \mathrm{Iso}(P,P')$ we see that the conditions for Theorem 4.6 are satisfied. There exists a C^∞ vector bundle $r: \mathcal{O} \to \mathrm{Iso}(P,P')$ which is an *open* subset of $E_{\mathbb{M}}(P,P')$. It follows that for $\bar{V}$ compact, there exists $\varepsilon_0 > 0$ such that for all $\varepsilon < \varepsilon_0$, every g_ε satisfying (4.13) is in $C^\infty(V; \mathcal{O})$. Fix $0 < \varepsilon < \varepsilon_0$ and g_ε satisfying (4.13). Then $\theta = r \circ g_\varepsilon$ is a global, C^∞ section of $\mathrm{Iso}(P,P')$. This proves Theorem 4.4, when V is compact.

V.5. THE PATCHING PROBLEM: COMPACT DOMAINS

The proof of Theorem 2.4 is completed below. The consequence of Proposition 3.1 is that locally (A,Φ) is gauge equivalent to a smooth configuration. To prove Theorem 2.4, it is necessary to patch these smooth gauges together. We consider this situation in a broader geometric framework.

PROPOSITION 5.1. *Let* (A,Φ) *and* $V \subseteq U \subseteq M$ *be as in Theorem* 2.4. *There exists a* C^∞ *principal* G*-bundle* $P'|_V$ *and a* C^0 *isomorphism* $\theta: P|_V \to P'|_V$ *such that* (A_θ, Φ_θ) *are respectively a smooth connection on* $P'|_V$ *and a smooth section of* $E'|_V = E\theta|_V$. (See Definitions 4.1-2.)

Assuming Proposition 5.1, the proof of Theorem 2.4 is straightforward: By Theorem 4.4, there exists a C^∞ bundle isomorphism $\theta': P'|_V \to P|_V$. The configuration

$$((A_\theta)_{\theta'}, (\Phi_\theta)_{\theta'})$$

is respectively a smooth connection on $E|_V$ and a smooth section of $E|_V$. But $g \equiv \theta' \circ \theta : P|_V \to P|_V$ is a global gauge transformation. Thus $(A_g, \Phi_g) = (gAg^{-1} + gdg^{-1}, g\Phi g^{-1})$ are smooth in V and gauge related to (A,Φ) as claimed. (See Chapter IV.6 for an explicit example.)

Proof of Proposition 5.1. Since the closure $\bar{V}$ of V is compact, every open cover of V has a finite subcover. Due to Proposition 3.1, V is covered by a finite set of balls $\{B_i\}_{i\in\Lambda}$ such that the following is true: In each B_i there exists a C^0 map $g_i: B_i \to G$ such that

$$(\tilde{A}_i = g_i A_i g_i^{-1} + g_i dg_i^{-1}, \quad \tilde{\Phi}_i = \rho(g_i)\Phi_i) \tag{5.1}$$

is smooth. Here we represent (A,Φ) by $\{(A_i,\Phi_i)\}_{i\in\Lambda}$ as discussed in the previous section.

The bundle $P|_V$ has transition functions $h_{ij}: B_i \cap B_j \to G$ which are defined for $B_i \cap B_j \neq \emptyset$. With respect to the cover $\{B_i\}$, define a principal G-bundle P' by the transition functions

$$h'_{ij} = g_i h_{ij} g_j^{-1} \quad . \tag{5.2}$$

Let E' be the vector bundle associated to P' by the representation ρ.

LEMMA 5.2. *The configuration* $(\tilde{A},\tilde{\Phi})$ *defined by the set* $\{(\tilde{A}_i,\tilde{\Phi}_i)\}_{i\in\Lambda}$ *is respectively a connection on* P' *and a section of* E'.

Proof. It suffices to check how $(\tilde{A}_i,\tilde{\Phi}_i)$ relate to $(\tilde{A}_j,\tilde{\Phi}_j)$ in $B_i \cap B_j$. Then,

$$\tilde{\Phi}_i = \rho(g_i)\Phi_i = \rho(g_i h_{ij})\Phi_j = \rho(g_i h_{ij} g_j^{-1})\tilde{\Phi}_j$$

$$= \rho(h'_{ij})\tilde{\Phi}_j \quad , \qquad x \in B_i \cap B_j \, . \qquad (5.3)$$

Similarly

$$\tilde{A}_i = h'_{ij}\tilde{A}_j h'^{-1}_{ij} + h'_{ij} dh'_{ij} \quad , \qquad x \in B_i \cap B_j \, . \qquad (5.4)$$

LEMMA 5.3. *The principal bundle* P' *is smooth.*

Proof. It is sufficient to check that the transition functions h'_{ij} are smooth. This fact follows directly from (5.4); the proof is similar to the proof of Proposition IV.6.2. So we omit it.

The relevant consequences of Lemma 5.3 are first that E' is also smooth, and second that $(\tilde{A},\tilde{\Phi})$ are a smooth connection on P' and a smooth section of E' respectively. The set $\{g_i\}_{i\in\Lambda}$ are by construction the C^0 isomorphism from P to P'. This completes the proof of Proposition 5.1.

V.6. SMOOTHNESS OF SOLUTIONS ON $\mathbb{R}^d$

Before finishing the proof of Theorem 1.1 we summarize what has been established in Sections 2-5. Suppose that (A,Φ) is locally square integrable with locally square integrable first derivatives on $\mathbb{R}^d$, $d = 2,3$ and that equation (2.2) is satisfied in every open set $U \subset \mathbb{R}^d$. Then by Proposition 3.1, there exists a locally finite cover of $\mathbb{R}^d$ by open balls $\{B_i\}_{i\in\Lambda}$ and C^0 gauge transformations $g_i : B_i \to G$ such that in each B_i,

$$\{(\tilde{A}_i = g_i A g_i^{-1} + g_i dg_i^{-1}, \quad \tilde{\Phi}_i = g_i \Phi g_i^{-1})\}_{i\in\Lambda} \qquad (6.1)$$

are smooth. Arguing as in §5, the set $\{\tilde{A}_i\}_{i\in\Lambda}$ define a C^∞ connection

$\tilde{A}$ on a smooth, principal G bundle P'. The set $\{\tilde{\Phi}_i\}$ define a C^∞ section of the associated vector bundle E'. Further, there exists a C^0 isomorphism $\theta: \mathbb{R}^d \times G \to P'$ and the induced isomorphism $\theta: \mathbb{R}^d \times \mathbb{L} \to E'$. Hence Theorem 1.1 follows from the existence of a C^∞ isomorphism $\theta': P' \to \mathbb{R}^d \times G$. This we now prove.

THEOREM 6.1. *Let* G *be a compact, connected Lie group. Let* $p': P' \to \mathbb{R}^d$ *be a smooth principal* G*-bundle which is* C^0 *isomorphic to the product bundle* $P = \mathbb{R}^3 \times G$. *Then* P' *is* C^∞ *isomorphic to* P.

<u>Remark</u>. In fact, every C^0 principal G-bundle on $\mathbb{R}^d$ is C^0 isomorphic to $\mathbb{R}^3 \times G$. See, e.g., Steenrod [4], Corollary I.11.6.

<u>Proof</u>. Define $B_N = \{x \in \mathbb{R}^d: |x| < N\}$, $N = 1,2,\ldots$. By Theorem 4.4, there exists a smooth isomorphism $\theta_N: P'|_{B_N} \to P|_{B_N}$.

It follows that on B_N, $g_N = \theta_{N+1}\theta_N^{-1}$ is a gauge transform, a section of $\mathrm{Iso}(P|_{B_N}; P|_{B_N})$. This is nothing more than a C^∞ map $g_N: B_N \to G$.

LEMMA 6.2. *Let* $R > 0$. *Suppose that* $g: B_R \to G$ *is smooth. There exists a smooth extension* $\hat{g}$ *of* g *such that*

$$(1) \quad \hat{g} . \mathbb{R}^d \to G$$

$$(2) \quad \hat{g} = g \quad \text{on} \quad B$$

$$(3) \quad \hat{g} = 1 \quad \text{for} \quad x \in \mathbb{R}^d \setminus B_{R+1/2} \quad . \tag{6.2}$$

<u>Proof</u>. Because G is connected, there exists a C^∞ map $h(t): [0,1] \to G$ such that

$$(1) \quad h(0) = 1$$

$$(2) \quad h(1) = g^{-1}(x=0) \quad . \tag{6.3}$$

We can also find a smooth map $\psi: \mathbb{R}^d \setminus B_R$ satisfying

$$(1)\quad \psi(x) = x \quad \text{if} \quad |x| \le R \quad ,$$

$$(2)\quad \psi(x) = 0 \quad \text{if} \quad |x| \ge R + 1/2 \quad . \tag{6.4}$$

Define $\hat{g}_N(x)$ to be

$$\hat{g}(x) = g(\psi(x))h(|x-\psi(x)|^2/|x|^2) \quad . \tag{6.5}$$

Define smooth isomorphisms $\hat{\theta}_N: P'|_{B_N} \to P|_{B_N}$ recursively as follows: Set

$$(1)\quad \hat{\theta}_1 = \theta_1$$

$$(2)\quad \hat{\theta}_2 = \hat{g}_1^{-1}\theta_2$$

$$(3)\quad \hat{\theta}_{N+1} = \hat{g}_N^{-1}\theta_{N+1}, \quad N = 3,\ldots,\infty \tag{6.6}$$

Here $\hat{g}_N$ is the smooth extension of $g_N = \theta_{N+1}\hat{\theta}_N^{-1}: B_N \to G$ that is given by (6.5) and Lemma 6.2.

By construction, the restriction

$$\hat{\theta}_N|_{B_{N-1}} = \hat{\theta}_{N-1} \quad . \tag{6.7}$$

Hence the set $\{\hat{\theta}_N: B_N \to G\}_{N=1}^{\infty}$ defines a global, smooth isomorphism $\theta': P' \to \mathbb{R}^3 \times G$ by

$$\theta'|_{B_N} = \hat{\theta}_N \tag{6.8}$$

The existence of θ' proves Theorem 6.1.

V.7. SOME A PRIORI ESTIMATES ON $\mathbb{R}^2$, $\mathbb{R}^3$

Suppose $u \in C^\infty(\mathbb{R}^d; E)$ and satisfies

$$-\nabla_A^2 u = v \tag{7.1}$$

for some v. Here A is a smooth connection on E (cf. VI §5 for notation). In this section we will derive a priori bounds (based on v) on the L_p and L_p^1 norms of $|u|$. For $p > d$ the L_p^1 bound implies uniform decay of $|u|$ as $|x|$ tends to infinity (cf. Ch. III, Proposition III.7.5). Loosely speaking, the proofs contained in this section involve no more than justifying the use of integration by parts and some Sobolev inequalities (cf. Ch. VI, §§1 and 2).

For a nonlinear problem, v in (7.1) is a function of u and often a priori L_p estimates on v cannot be established. Below we consider a special case.

THEOREM 7.1. *Take* $d = 2,3$. *Let* A *be a smooth connection on* E *with* L_2 *curvature* F_A. *Let* $u \in C^\infty(\mathbb{R}^d; E) \cap L_2(\mathbb{R}^d; E)$. *Let* $\Lambda(\cdot)$ *be an* $n \times n$ *matrix with* L_2 *coefficients and* $v \in L_2(\mathbb{R}^2; E)$. *Suppose that*

$$-\nabla_A^2 u = \Lambda(u) + v \quad . \tag{7.2}$$

Then

$$\text{(a)} \quad |\nabla_A u| \in H_1(\mathbb{R}^d) \tag{7.3a}$$

$$\text{(b)} \quad |u| \in L_p^2(\mathbb{R}^d) \quad , \tag{7.3b}$$

where $p \in [2,6]$ if $d = 3$ and $2 \le p < \infty$ if $d = 2$.

The proof of Theorem 7.1 is simplified by first proving the following proposition. The proposition below gives an a priori bound on $\|u\|_\infty$ for a solution u to (7.2) which is independent of the L_2 norm of curvature F_A.

PROPOSITION 7.2. *Take* $d=2,3$. *Let* A *be a smooth connection on* E *(with no restriction on the curvature). Let* $\Lambda(\cdot)$, v *be as in Theorem* 7.1 *and together with* $u \in C^\infty(\mathbb{R}^d; E) \cap L^2(\mathbb{R}^d; E)$ *satisfy the inequality*

$$-(u,\nabla_A^2 u) \le (u,\Lambda(u)) + (u,v) \quad . \tag{7.4}$$

Then

$$\|\nabla_A u\|_{L_1}^2 \le c_1 \|\Lambda\|_{L_2}^4 \; \|u\|_{L_2}^2 + 2\langle u,v\rangle_{L_2} \tag{7.5}$$

and

$$\|u\|_\infty \le c_2 \left(\|u\|_{L_2} + \|v\|_{L_2} + \left(\|\nabla_A u\|_{L_2} + \|u\|_{L_2} \right) \|\Lambda\|_{L_2}^2 \right) \quad . \tag{7.6}$$

Here, $c_1, c_2 < \infty$ *are independent of* A.

Proof. The philosophy is to control the L_2 norm of $|\nabla_A u|$ in a ball of radius R and obtaining estimates which are R-independent. This is done by introducing a cut off. As usual, we use the function $b_R(x)$ defined in Chapter VI.3 as a cut off. The important identity is

$$\begin{aligned} \nabla_{A_i}\nabla_{A_j} b_R u \; &= \; b_R \nabla_{A_i}\nabla_{A_j} u + (\nabla_i b_R)\nabla_{A_j} u \\ &\quad + (\nabla_j b_R)\nabla_{A_j} u + u\nabla_i\nabla_j b_R \quad . \end{aligned} \tag{7.7}$$

Multiplying both sides of (7.4) by b_R^2 and using the above identity, we obtain

$$\begin{aligned} -(u_R,\nabla_A^2 u_R) &\le (u_R,\Lambda(u_R)) + (u_R,v_R) \\ &\quad - 2(\nabla_i b_R)(u_R,\nabla_{A_i} u) - (\Delta b_R)(u,u_R) \quad . \end{aligned} \tag{7.8}$$

Here $u_R = b_R u$ and $v_R = b_R v$. The element u_R is compactly supported so there is no problem with integrating (7.8) over $\mathbb{R}^d$ and then integrating the left hand side by parts. The result of these two operations is

$$\|\nabla_A u_R\|_{L_2}^2 \le \|\Lambda|u_R|^2\|_{L_1} + \langle u_R, v_R\rangle_{L_2} + \frac{\|\Delta b\|_\infty}{R^2} \|u\|_{L_2}^2 \quad . \qquad (7.9)$$

In deriving (7.8), we have taken $R \to \infty$ in the terms on the right hand side which are finite by assumption. We have used the scaling properties of b_R as given in Lemma VI.3.2. If it is assumed that $|\Lambda| \in L_\infty$ then we are essentially done, and the dominated convergence theorem gives

$$\|\nabla_A u\|_{L_2}^2 \le \|\Lambda\|_\infty \|u\|_{L_2}^2 + \langle u, v\rangle_{L_2} \quad . \qquad (7.10)$$

Under the assumption that $|\Lambda| \in L_2$, we must isolate the R dependence of the first term on the right hand side of (7.9). From Hölder's inequality

$$\||\Lambda||u_R|^2\|_{L_1} \le \|\Lambda\|_{L_2} \|u\|_{L_2}^{1/2} \|u\|_{L_6}^{3/2} \quad . \qquad (7.11)$$

Next use a Sobolev inequality (Corollary VI.2.6) to write

$$\|u_R\|_{L_6}^{3/2} \le c^{3/4} \|\nabla |u_R|\|_{H_1}^{3/2} \quad . \qquad (7.12)$$

and use Kato's inequality to obtain

$$\|u_R\|_{L_6}^{3/2} \le c^{3/4} \left(\|\nabla_A u_R\|_{L_2}^2 + \|u\|_{L_2}^2\right)^{3/4} \quad . \qquad (7.13)$$

Substitute (7.13) into (7.11) and use the resulting inequality in (7.9). This gives

$$\|\nabla_A u_R\|_{L_2}^2 \le \left(c^{3/4} \|\Lambda\|_{L_2} \|u\|_{L_2}^{1/2}\right) \left(\|\nabla_A u_R\|_{L_2}^2 + \|u\|_{L_2}^2\right)^{3/2}$$
$$+ \langle u_R, v_R\rangle_{L_2} + \frac{\|\Delta b\|_\infty}{R^2} \|u\|_{L_2}^2 \quad . \qquad (7.14)$$

Equation (7.14) implies a bound on $\|\nabla_A u_R\|_{L_2}$ which is independent of R. This is made manifest by using the inequality

$$ab^{3/2} \leq 4a^2b + \frac{1}{4} b^2 \leq 64a^4 + \frac{1}{2} b^2 \quad . \tag{7.15}$$

Thus

$$\|\nabla_A u_R\|_{L_2}^2 \leq 128c^2 \|\Lambda\|_{L_2}^4 \|u\|_{L_2}^2 + 2\langle u_R, v_R\rangle_{L_2} + \frac{1}{2} \|u\|_{L_2}^2 + \frac{2}{R^2} \|\Lambda b\|_\infty \|u\|_{L_2}^2 \quad . \tag{7.16}$$

By the triangle inequality,

$$\|b_R \nabla_A u\|_{L_2}^2 \leq \left(\frac{R+1}{R}\right) \|\nabla_A u_R\|^2 + \left(\frac{1+R}{R^2}\right) \|\nabla b\|_\infty^2 \|u\|_{L_2} \quad . \tag{7.17}$$

Here, Lemma VI.3.2 has been used. Substituting (7.17) into (7.16) and using the dominated convergence theorem gives (7.5).

To prove (7.6), we return to (7.8) and use the inequality

$$-\frac{1}{2} \Delta |u_R|^2 = -(u_R, \nabla_A^2 u_R) - |\nabla_A u_R|^2 \leq -(u_R, \nabla_A^2 u_R) \quad . \tag{7.18}$$

Hence

$$(-\Delta + 1)|u_R|^2 \leq 2|u_R|^2 |\Lambda| + 2|u_R||v_R| + |u_R|^2$$

$$+ 4|\nabla b_R||u_R||\nabla_A u| + 2|\Delta b_R||u_R||u| \quad . \tag{7.19}$$

In order to use (7.19) to derive an integral inequality for $|u_R(x)|^2$, it is necessary to introduce the Green's function for the operator $(-\Delta + 1)$. This Green's function is denoted $G(x-y)$ and aside from the properties listed in Proposition VI.4.8, $G \in L_p$ for $1 \leq p < d/d-2$ for $d = 2,3$. That $G \in L_1$ follows from equation (VI.4.41). That $G \in L_p$, $2 \leq p < d/d-2$ follows from (VI.4.41) and the Hausdorff-Young inequality.

Define a comparison function $s_R(x)$ by,

$$s_R(x) = \int dy\, G(x-y) \Big\{ 2|u_R|^2|\Lambda| + 2|u_R||v_R| + |u_R|^2 + 4|\nabla b_R||u_R||\nabla_A u| + 2|\Delta b_R||u_R||u| \Big\}(y) \quad . \tag{7.20}$$

By the maximum principle, Proposition VI.3.1,

$$|u_R|^2(x) \le s_R(x) \quad . \tag{7.21}$$

For notational convenience, set $t_R = \|u_R\|_\infty$. Then, using Hölder's inequality,

$$s_R(x) \le 2t_R^{3/2}\|\Lambda\|_{L_2}\|u\|_{L_6}^{1/2}\|G\|_{L_{12/5}} + 2t_R\|v\|_{L_2}\|G\|_{L_2} + 2t_R\|u\|_{L_2}\left(1 + \|\Delta b\|_\infty \frac{1}{R^2}\right)\|G\|_{L_2} + 4t_R\frac{\|\nabla b\|_\infty}{R}\|\nabla_A u\|_{L_2}\|G\|_{L_2} \tag{7.22}$$

Here we have used the scaling properties of $b_R(x)$ (see Lemma VI.3.2). Using (7.21) and (7.22) we obtain

$$t_R^2 \le \alpha_R t_R^{3/2} + \beta_R t_R \quad ,$$

where

$$\alpha_R = 2\|\Lambda\|_{L_2}\|u\|_{L_6}^{1/2}\|G\|_{L_{12/5}} \quad ,$$

$$\beta_R = 2\|G\|_{L_2}\left(\|v\|_{L_2} + \|u\|_{L_2}\left(1 + \frac{1}{R^2}\|\Delta b\|_\infty\right) + \frac{2}{R}\|\nabla b\|_\infty\|\nabla_A u\|_{L_2}\right) \quad . \tag{7.23}$$

Thus from (7.23)

$$t_R \le 2(\beta_R + 2\alpha_R^2) \quad . \tag{7.24}$$

Since t_R is an increasing function of R, we may take the $\liminf\ R\to\infty$ on the right hand side of (7.24). The result is

$$\sup_{|x|\le R} |u(x)| \le c_1'\left(\|v\|_{L_2} + \|u\|_{L_2} + \|\Lambda\|_{L_2}^2 \|u\|_{L_6}\right)$$

$$\le c_1\left(\|v\|_{L_2} + \|u\|_{L_2} + \|\Lambda\|_{L_2}^2 \left(\|\nabla_{A_0} u\|_{L_2} + \|u\|_{L_2}\right)\right). \qquad (7.25)$$

Here we have used a Sobolev inequality (Proposition VI.2.5) to bound $\|u\|_{L_6}$ by $\mathrm{const}\cdot(\|\nabla|u|\|_{L_2} + \|u\|_{L_2})$ and then Kato's inequality (Chapter VI.6) to bound $\|\nabla|u|\|_{L_2}$ by $\|\nabla_A u\|_{L_2}$. Since R is arbitrary in (7.25) we get (7.6) as claimed.

Proof of Theorem 7.1. Using (7.2) and (7.7) we derive the following equation for u_R:

$$-\nabla_A^2 u_R = \Lambda(u_R) + v_R - 2\nabla_i b_R \nabla_{A_i} \nabla u - (\Delta b_R)u \quad . \qquad (7.26)$$

Because $u \in L_\infty$, $\Lambda(u_R) \in L_2$ and

$$\|\Lambda(u_R)\|_{L_2} \le \|\Lambda\|_{L_2} \|u\|_\infty \quad .$$

Similarly,

$$\|\nabla_i b_R \nabla_{A_i} u\|_{L_2} \le \frac{\|\nabla b\|_\infty}{R} \|\nabla_A u\|_{L_2} < \infty$$

and

$$\|\Delta b_R u\|_{L_2} \le \frac{1}{R^2} \|\Delta b\|_\infty \|u\|_{L_2} \quad .$$

Thus, the right hand side of (7.26), which we denote as $\bar{v}_R$ is in L_2. The left hand side of (7.26) is compactly supported, so we square both sides and integrate over $\mathbb{R}^3$. The result of this operation is

$$\|\nabla_A^2 u_R\|_{L_2} = \|\bar{v}_R\|_{L_2}^2 \quad . \qquad (7.27)$$

On the other hand, an integration by parts relates $\|\nabla_A^2 u_R\|_{L_2}$ to the more useful quantity $\|\nabla_{A_i}\nabla_{A_j}u_R\|_{L_2}$:

$$\|\nabla_A^2 u_R\|_{L_2}^2 = \|\nabla_{A_i}\nabla_{A_j}u_R\|_{L_2}^2 + <\nabla_{A_i}\nabla_{A_j}u_R, F_{A_{ij}}(u_R)>$$

$$- <\nabla_{A_i}u_R, F_{A_{ij}}(\nabla_{A_j}u_R)> \quad . \tag{7.28}$$

(See Proposition VI.5.2 for $[\nabla_{A_i},\nabla_A^2]$.) Here we are considering $F_{A_{jk}}(\cdot)$ as an $n\times n$ matrix, a pointwise endomorphism of the vector space $\mathbb{L}$, as in Chapter VI.5.

From (7.28) and (7.27) we obtain

$$\|\nabla_{A_i}\nabla_{A_j}u_R\|_{L_2}^2 \le 2\|\bar{v}_R\|_{L_2}^2 + 4\|u\|_\infty^2\|F_A\|_{L_2}^2 + 2\||\nabla_A u|^2|F_A|\|_{L_1} \quad . \tag{7.29}$$

We remark on the similarity between (7.9) and (7.29). We treat the term $\||\nabla_A u|^2|F_A|\|_{L_1}$ in exactly the same way as the term $\||u|^2|\Lambda|\|_{L_1}$ was treated, cf. equations (7.11-6). The result is

$$\|\nabla_{A_1}\nabla_{A_j}u_R\|_{L_2}^2 \le 2\left\{2\|\bar{v}_R\|_{L_2}^2 + 4\|u\|_\infty^2\|F_A\|_{L_2}^2\|\nabla_A u\|_{L_2}^2 \right.$$

$$\left. + 256\, c_6^3\|F_A\|_{L_2}^4\|\nabla_A u\|_{L_2}^2\right\} \tag{7.30}$$

Using (7.7) to relate $\nabla_{A_i}\nabla_{A_j}u_R$ to $b_R\nabla_{A_i}\nabla_{A_j}u$ in (7.30) and then using the monotone convergence theorem, we obtain

$$\|\nabla_{A_i}\nabla_{A_j}u\|_{L_2} \le 2\left\{2\|v\|_{L_2} + \|\Lambda\|_{L_2}^2\|u\|^2 + 4\|F_A\|_{L_2}^2\|u\|^2\right.$$

$$\left. + \|\nabla_A u\|_{L_2}^2 + 256c^3\|F_A\|_{L_2}^4\|\nabla_A u\|_{L_2}^2\right\} \quad . \tag{7.31}$$

Equation (7.3a) of Theorem 1 follows from (7.31) by Kato's inequality (Chapter VI.6). Equation (7.3b) follows from (7.3a) by a Sobolev inequality, Proposition VI.2.5.

V.8. ESTIMATES FOR SOLUTIONS OF $\nabla_A^2 u = v$

We continue the notation of the previous section. In the last section we considered solutions u to (7.1) which were in L_2. In this section we will assume $u \in L_\infty$ but $\nabla_A u \in L_2$. That is, we allow the possibility that $|u| \to c \neq 0$ as $|x| \to \infty$.

THEOREM 8.1. *Let* $u \in L_\infty$ *and* $\nabla_A u \in L_2$. *Suppose that* $v \in L_2$ *and*

$$\nabla_A^2 u = v \quad . \tag{8.1}$$

Then $\nabla_{A_j} \nabla_{A_k} u \in L_2$, $|\nabla_A u| \in H_1$ *and*

$$\| |\nabla_A u| \|_{H_1} \leq c \left(\|f\|_{L_2}^2 + \|\nabla_A u\|_{L_2}^2 + \|f\|_{L_2}^4 \|\nabla_A u\|_{L_2}^2 + \|v\|_{L_2}^2 \right) \quad . \tag{8.2}$$

Proof. Define u_R as before. Using (7.7) we obtain

$$\nabla_A^2 u_R = b_R v + 2\nabla_j b_R \nabla_A u + \Delta b_R u \quad . \tag{8.3}$$

Squaring both sides of (8.3) and integrating over $\mathbb{R}^3$ gives

$$\|\nabla_A^2 u_R\|_{L_2} \leq \|v\|_{L_2} + \frac{2\kappa}{R} \|\nabla_A u\|_{L_2} + \|u\|_\infty \frac{\kappa'}{R^{1/2}} \quad , \tag{8.4}$$

where $\kappa = \|\nabla b\|_\infty$ and $\kappa' = \|\Delta b\|_{L_2}$ (see Lemma VI.3.2). To put (8.4) in a more usable form we use (7.28). This establishes that

$$\|\nabla_{A_j} \nabla_{A_k} u_R\|_{L_2} \leq 2 \left[\|v\|_{L_2} + \|F_A\|_{L_2} + \| |\nabla_A u_R|^2 |F_A| \|_{L_1}^{1/2} + c_2 R^{-1/2} \right] \tag{8.5}$$

Here c_2 is a finite constant, independent of $R > 1$. The only troublesome term in (8.5) is $\| |\nabla_A u_R|^2 |f| \|_{L_2}$. This is handled exactly as in the proof of Theorem 7.1 and we omit the details. The result is that

$$\|\nabla_{A_i}\nabla_{A_j} u_R\|_{L_2} \le c_1\left[\|v\|_{L_2} + \|f\|_{L_2} + \|f\|_{L_2}^2 \|\nabla_A u\|_{L_2}\right] + c_3 R^{-1/2} \tag{8.6}$$

where $c_1, c_3 < \infty$ are independent of $R > 1$. Using (7.7) again we obtain

$$\|b_R \nabla_{A_j}\nabla_{A_k} u\|_{L_2} \le c_1\left[\|v\|_{L_2} + \|f\|_{L_2} + \|f\|_{L_2}^2 \|\nabla_A u\|_{L_2} + c_4 R^{-1/2}\right]. \tag{8.7}$$

Here $c_4 < \infty$ and is independent of $R > 1$. The monotone convergence theorem implies that

$$\|\nabla_{A_i}\nabla_{A_j} u\|_{L_2} \le c_1\left[\|v\|_{L_2} + \|f\|_{L_2} + \|f\|_{L_2}^2 \|\nabla_A u\|_{L_2}\right]. \tag{8.8}$$

The remaining statements in Theorem 8.1 follow from Sobolev inequalities (Proposition VI.2.4) and Kato's inequality (Chapter VI, §6).

VI. ANALYTIC TOOLS

In this chapter we collect together some of the analytic techniques used in the earlier chapters. While much of this material will be familiar for mathematical analysts, we include a brief treatment, in an attempt to make this book self-contained. We assume the reader is familiar with the definitions of Banach and Hilbert spaces.

VI.1. SOBOLEV SPACES, CONVERGENCE, ETC.

Let us consider the spaces $C^k(\mathbb{R}^d)$ of functions on $\mathbb{R}^d$ with k continuous derivatives. Namely for $f \in C^k$ and any $s = (s_1, \ldots, s_d)$ with $|s| = \Sigma_{j=1}^d s_j$, the derivatives

$$(\nabla^{(s)} f)(x) = \left[\prod_{j=1}^{d} \left(\frac{\partial}{\partial x_j} \right)^{s_j} \right] f(x), \qquad 1 \leq 1 \leq k \ , \tag{1.1}$$

exist and are continuous. The space $C^\infty(\mathbb{R}^d)$ is defined as the intersection of $C^k(\mathbb{R}^d)$ for $k = 0,1,\ldots$. The subspace C_0^k of C^k consists of the k-times differentiable functions with compact support, i.e. functions $f(x) \in C_0^k$ that vanish identically for $|x| \equiv (\Sigma_k x_k^2)^{1/2}$ sufficiently large. Also $C_0^\infty \equiv \bigcap_k C_0^k$.

The basic norm we use is the L_p norm. Let $f \in C_0^\infty$. For $p < \infty$ define [e.g. 1]

$$\|f\|_{L_p} = (\int |f(x)|^p \, dx)^{1/p} \ , \tag{1.2}$$

and for $p = \infty$ define

$$\|f\|_{L_\infty} = \sup_x |f(x)| \ . \tag{1.3}$$

The L_p spaces, $p<\infty$, are the Banach space completions of C_0^∞ in these norms.

The Sobolev spaces L_p^k are subspaces of the L_p spaces such that the functions f have k derivatives in L_p.

DEFINITION 1.1. *The Sobolev space* $L_p^k(\mathbb{R}^d)$ *is the Banach space completion of* $C_0^\infty(\mathbb{R}^d)$ *in the norm*

$$\|f\|_{k,p} = \left(\sum_{0\le|s|\le k} \|\nabla^{(s)} f\|_{L_p}^p \right)^{1/p} . \tag{1.4}$$

If $p=2$, *set*

$$H_k = L_2^k . \tag{1.5}$$

The space H_k is a Hilbert space with the natural inner product

$$\langle f,g\rangle_{H_k} = \sum_{0\le|s|\le k} \langle \nabla^{(s)} f, \nabla^{(s)} g\rangle_{L_2} . \tag{1.6}$$

For $d\ge 3$, it is also convenient to define H' by the norm

$$\|f\|_{H'} = \|\nabla f\|_{L_2} . \tag{1.7}$$

Several further generalizations of these spaces occur. First one may restrict $\mathbb{R}^d$ to a bounded open domain V with smooth boundary $\partial\bar{V}$.

DEFINITION 1.2. *The spaces* $L_p^k(V)$ *and* $L_{p(0)}^k(V)$ *are the completions of* $C^\infty(\bar{V})$, $C_0^\infty(V)$ *in the norm*

$$\|f\|_{k,p} = \sum_{0\le|s|\le k} \|\chi_V \nabla^{(s)} f\|_{L_p} , \tag{1.8}$$

where

$$\chi_V(x) = \begin{cases} 1 & x \in V \\ 0 & x \notin V \end{cases} .$$

A second generalization allows f to take values in a space of sections of differential forms, or more generally in sections of a vector bundle E over a manifold M; cf. Chapter V.2.

We now state some properties of L_p^k functions.

PROPOSITION 1.3. *Let* $f \in L_p(\mathbb{R}^d)$, $1 < p < \infty$ *and* χ_R *be the characteristic function of the set* $\{x \in \mathbb{R}^3 \mid |x| \leq R\}$. *Then*

$$\lim_{R\to\infty} \|(1 - \chi_R) f\|_{L_p} \to 0 . \tag{1.9}$$

PROPOSITION 1.4. *Let* $f \in L_p^1(\mathbb{R}^d)$, $1 \leq p < \infty$. *Then*

$$|f| \in L_p^1(\mathbb{R}^d) .$$

PROPOSITION 1.5. *Let* $V \subset \mathbb{R}^d$ *be a bounded domain with smooth boundary* $\partial\bar{V}$. *If* $f \in L_2^k(V)$ *then* $f \in L_2^{k-1}(\partial\bar{V})$. *Further, if* $f \in L_{2(0)}^k(V)$ *then for* $|s| \leq k-1$,

$$\nabla^{(s)} f \Big|_{\partial\bar{V}} = 0 \qquad \textit{almost everywhere.}$$

PROPOSITION 1.6. *Translation is continuous on* L_p^k, $p < \infty$. *In other words, for all* $f \in L_p^k$,

$$\lim_{a\to 0} \|f - T_a f\|_{L_p^k} = 0 \tag{1.10}$$

where

$$(T_a f)(x) = f(x-a) .$$

The proofs of these propositions are facilitated by introducing the notion of a transformation T from a Banach space X to a Banach space

Y. Then T is continuous if whenever $\|f_n - f\|_X \to 0$, then also $\| T(f_n) - T(f) \|_Y \to 0$.

DEFINITION 1.7. *A transformation* T *from a Banach space* X *to a Banach space* Y *is* linear *if for all* $f, g \in X$ *and* $\lambda \in \mathbb{R}$ *(or* $\lambda \in \mathbb{C}$ *if* X,Y *are complex)*

$$T(f + \lambda g) = T(f) + \lambda T(g) \quad . \tag{1.11}$$

A linear transformation is bounded *with norm* $\|T\|_{Y,X}$ *if*

$$\|T\|_{X,Y} \equiv \sup_{\|f\|_X=1} \|Tf\|_Y < \infty \,. \tag{1.12}$$

In case X = Y, *we write*

$$\|T\|_X = \|T\|_{X,X} \quad .$$

Remark. If T is linear, then T is continuous if and only if T is bounded.

DEFINITION 1.8. *A family* $\{T_n\}$ *of transformations* $T_n : X \to Y$ *converges (strongly) to* T *if for every* $f \in X$,

$$\lim_{n\to\infty} \|T_n(f) - T(f)\|_Y \to 0 \quad . \tag{1.13}$$

PROPOSITION 1.9. (3 ε argument) *Let* T_n *be a family of linear transformations between Banach spaces* X *and* Y *which are uniformly bounded, i.e.*

$$\|T_n\|_{X,Y} \le M \quad , \quad \text{all } n \quad , \tag{1.14}$$

and convergent for f *in a dense subset* $X_0 \subset X$, *i.e.*

$$\|T_n f - Tf\| \to 0 \quad , \quad f \in X_0 \quad . \tag{1.15}$$

Then $T_n f \to Tf$ *for all* $f \in X$, *and*

$$\|T\|_{Y,X} \le M \quad . \tag{1.16}$$

Remark. If $\{T_n\}$ are nonlinear, but are continuous, uniformly in n, the 3ε argument also holds.

Proof. Given $f \in X$, there exists $f_0 \in X_0$ with $\|f - f_0\|_X < \varepsilon$. Then

$$\begin{aligned} \|T_n f - T_m f\|_Y &= \|T_n(f - f_0) + (T_n - T_m)f_0 + T_m(f_0 - f)\|_Y \\ &\le 2M\|f - f_0\|_X + \|(T_n - T_m)f_0\|_Y \\ &\le 2M\varepsilon + \|(T_n - T_m)f_0\|_Y \quad . \end{aligned}$$

Choose n,m sufficiently large so $\|(T_n - T_m)f_0\| < \varepsilon$. Thus $T_n f \to$ limit for all $f \in X$. Clearly the limit is linear in f and it defines Tf. But

$$\|Tf\|_Y = \|(T - T_n)f + T_n f\|_Y \le \overline{\lim_n}\, \|(T - T_n)f\|_Y + M\|f\|_X \tag{1.17}$$

Armed with the 3ε argument, we prove Propositions 1.3-1.6.

Proof of Proposition 1.3. Define a transformation $T(R): L_p \to L_p$ by

$$T(R)f = (1 - \chi_R)f \quad , \qquad R \ge 0 \; .$$

Then $T(R)$ is linear and $\|T(R)\|_{L_p} \le 1$. The set $C_0^\infty(\mathbb{R}^d)$ is dense in L_p, $p < \infty$. Hence by the 3ε argument it is sufficient to show

$$\lim_{R\to\infty} T(R)f = 0 \tag{1.18}$$

for all $f \in C_0^\infty(\mathbb{R}^d)$. But $T(R)f \equiv 0$ for R sufficiently large.

Proof of Proposition 1.4. Define a transformation $T: C_0^\infty(\mathbb{R}^d) \to C_0^0(\mathbb{R}^d)$ by

$$(Tf)(x) = f_+(x) = \begin{cases} f(x), & f(x) \geq 0 \\ 0 & f(x) \leq 0 \end{cases} . \tag{1.19}$$

The function f_+ is piecewise continuous; its derivative is bounded and defined almost everywhere. Hence T maps $C_0^\infty(\mathbb{R}^d) \to L_2^1(\mathbb{R}^d)$, and for $f \in C_0^\infty$

$$\|Tf\|_{L_p^1} \leq \|f\|_{L_p^1} . \tag{1.20}$$

Hence T extends by continuity to L_p^1. Likewise f_-, $|f|$ are in L_p^1.

Proof of Proposition 1.5. It is enough to prove the proposition for $f \in L_2^1(V)$ since by substituting $f \to \nabla^{(s)} f$ we reduce to this case.

Let η be the volume form for $\partial\bar{V}$. That is, we consider $\partial\bar{V}$ to be a smooth manifold of dimension $d-1$. Locally, η is the zero set of a C^∞ function g such that $dg \neq 0$. Then locally

$$\eta = \frac{1}{(d-1)!} \frac{1}{|\nabla g|} \varepsilon^{ki_1 \ldots i_{d-1}} \nabla_k g dx^{i_1} \wedge \ldots \wedge dx^{i_{d-1}} \tag{1.21}$$

where $\varepsilon^{i_1 \ldots i_d}$ is completely antisymmetric with $\varepsilon^{12\ldots d} = 1$. We extend η to a smooth $(d-1)$-form $\hat{\eta}$ in the interior of V.

Define a transformation $T: C^\infty(V) \to C^\infty(\partial\bar{V})$ by taking Tf to be the restriction of f to $\partial\bar{V}$. For $f \in C^\infty(V)$,

$$\|Tf\|^2_{L_2(\partial\bar{V})} = \int_{\partial\bar{V}} |f|^2 \eta . \tag{1.22}$$

By Stokes' Theorem,

$$\|Tf\|^2_{L_2(\partial\bar{V})} = \int_V \left\{ 2|f| d|f| \wedge \hat{\eta} + |f|^2 d\hat{\eta} \right\}$$

$$\leq C_1 \left(\|f\|_{L_2} \|\nabla f\|_{L_2} + \|f\|^2_{L_2} \right) \leq C_2 \|f\|^2_{H_1(V)} \tag{1.23}$$

where $C_1, C_2 < \infty$ depend only on the particular extension $\hat{\eta}$ of η. Thus T is uniformly bounded. By the 3ε argument, T is a bounded, linear transformation from $L_2^1(V)$ to $L_2(\partial\bar{V})$.

Since $Tf = 0$ for all $f \in C_0^\infty(V)$, we conclude that $Tf = 0$ for all $f \in L^1_{2(0)}(V)$. This follows from the 3ε argument as well.

Proof of Proposition 1.6. Since the L_p^k norms are translation invariant, $\|T_a\|_{L_p^k} = 1$. By the 3ε argument, it is enough to consider $f \in C_0^\infty(\mathbb{R}^d)$. For such f,

$$|f(x) - f(x-a)| = \left| \int_0^1 dt\; a^j \nabla_j f(x-ta) \right|$$

$$\leq |a| \sup_{|y| \leq a} |(\nabla f)(x+y)| \quad .$$

Thus (1.10) follows.

VI.2. ELEMENTARY INEQUALITIES

The best known L_p inequality is the Schwarz inequality. A generalization is

PROPOSITION 2.1. (Hölder's Inequality) *Let* $1 \leq p \leq \infty$, *with*

$$\frac{1}{p} + \frac{1}{p'} = 1 \quad . \tag{2.1}$$

Then for $a, b \geq 0$ *any* f, g

$$a^{1/p} b^{1/p'} \leq \frac{1}{p} a + \frac{1}{p'} b \tag{2.2a}$$

and for

$$\|fg\|_{L_1} \leq \|f\|_{L_p} \|g\|_{L_{p'}} \quad . \tag{2.2b}$$

The indices (p,p') *satisfying* (2.1) *are called* dual *indices.*

Proof. Note that for $a > 0$, $p \geq 1$,

$$a \leq \frac{1}{p} a^p + \frac{1}{p'} \quad ,$$

since the only critical point of $f(a) = a - \frac{1}{p} a^p$ on $[0,\infty)$ is a maximum at $a = 1$. Now substitute $a \to (a/b)^{1/p}$ to recover (2.2a). Assume $\| f\|_{L_p} = \|g\|_{L_{p'}} = 1$. Then integration of (2.2a) with $a = |f|^p$, $b = |g|^{p'}$, yields (2.2b). The general case follows by homogeneity in f,g.

Next we introduce the Fourier transform on $\mathbb{R}^d$. Let

$$f(x) = \frac{1}{(2\pi)^{d/2}} \int_{\mathbb{R}^d} \tilde{f}(k) e^{-ikx} \, dk \quad , \tag{2.3}$$

$$\tilde{f}(k) = \frac{1}{(2\pi)^{d/2}} \int_{\mathbb{R}^d} f(x) e^{ikx} \, dx \quad . \tag{2.4}$$

The transformation $f \to \tilde{f}$ is unitary on $L_2(\mathbb{R}^d)$, i.e.

$$\langle f,f\rangle_{L_2} = \langle \tilde{f},\tilde{f}\rangle_{L_2} \quad . \tag{2.5}$$

More generally, the Fourier transform is a norm-decreasing transformation from L_p, $1 \leq p \leq 2$ to $L_{p/(p-1)}$:

PROPOSITION 2.2. (Hausdorff-Young Inequality) *Let* $2 \leq p$, *and* $1/p + 1/p' = 1$. *Then*

$$\|f\|_{L_p} \leq \|\tilde{f}\|_{L_{p'}} \quad . \tag{2.6}$$

Remark. For $p = 2$, equality is the statement (2.5). For $p = \infty$, taking absolute values in (2.3) yields

$$\|f\|_{L_\infty} \leq (2\pi)^{-d/2} \|\tilde{f}\|_{L_1} \leq \|\tilde{f}\|_{L_1}$$

Having established the extreme cases $p = 2, \infty$, the general case $2 < p < \infty$ follows from an interpolation theorem which we do not prove; see [10].

These two inequalities provide basic tools for many estimates. As an illustration of the usefulness of these two elementary inequalities, we establish a relation in $d = 2$ between

$$\|f\|_{H_1}^2 = \|f\|_{L_2}^2 + \|\nabla f\|_{L_2}^2 \geq \|f\|_{L_2}^2 \qquad \text{and} \qquad \|f\|_{L_p} .$$

PROPOSITION 2.3. *Let* $d = 2$ *and* $p > 2$. *Then* $H_1 \subset L_p$, *and*

$$\|f\|_{L_p} \leq \left[\pi \left(\frac{p-2}{2}\right)\right]^{\left(\frac{p-2}{2p}\right)} \|f\|_{H_1} . \tag{2.7}$$

Hence there exist c, c' *such that for all* $f \in H_1$,

$$\int_{\mathbb{R}^2} |e^f - 1 - f| dx \leq c \exp\left[c \|f\|_{H_1}^2\right] . \tag{2.8}$$

Proof. First remark that (2.8) follows from (2.7) by expanding $\exp f$ in a power series and using (2.7) on each term. By Stirlings formula, there are constants c_1, c_2 such that $(c_1 n)^n \leq n! \leq (c_2 n)^n$. Summing these bounds yields (2.8).

To establish (2.7), use Hölder's inequality and the Hausdorff-Young inequality as follows: For any $\alpha > 0$,

$$\begin{aligned} \left(\int |f|^p \, dx\right)^{1/p} &= \|f\|_{L_p} \leq \|\tilde{f}\|_{L_{p'}} \\ &= \left(\int |\tilde{f}(k)|^{p'} (1+k^2)^{\alpha} (1+k^2)^{-\alpha} \, dk\right)^{1/p'} \\ &\leq \|\tilde{f}^{p'} (1+k^2)^{\alpha}\|_{L_r}^{1/p'} \left(\int (1+k^2)^{-\alpha r'} \, dk\right)^{1/(p'r')} . \end{aligned} \tag{2.9}$$

Choose r and α so that $r\alpha = 1$, $rp' = 2$. Thus

$$\int |f|^p \, dx = \|f\|_{L_p}^p \le \left(\int (1+k^2)^{-\alpha r'} \, dk\right)^{p/(p'r')} \|f\|_{H_1}^{2p/(rp')} . \quad (2.10)$$

The choice of r and α entails

$$\alpha r' = \frac{p}{p-2} , \qquad \text{and} \qquad \frac{p}{p'r'} = \left(\frac{p-2}{2}\right) . \quad (2.11)$$

Thus

$$\int |f|^p \, dx \le \|f\|_{H_1}^p \left(\int (1+k^2)^{-1-2/(p-2)} \, dk\right)^{\left(\frac{p-2}{2}\right)} . \quad (2.12)$$

Note that for $\varepsilon > 0$,

$$\int_0^\infty \left(\frac{1}{t+1}\right)^{1+\varepsilon} dt = \frac{1}{\varepsilon} ,$$

and

$$\left(\int \left(\frac{1}{1+k^2}\right)^{1+\frac{2}{p-2}} dk\right)^{\left(\frac{p-2}{2}\right)} = \left[\pi \left(\frac{p-2}{2}\right)\right]^{\left(\frac{p-2}{2}\right)} ,$$

as desired.

The argument above keeps careful track of the growth of the constant in (2.7) as $p \to \infty$. The bound also shows that there exists $\varepsilon > 0$, such that for all $f \in H_1$, $\exp(\varepsilon f^2) - 1 \in L_1$. In dimension d, other inequalities like (2.7) hold.

PROPOSITION 2.4. (Sobolev Inequalities) *Let*

$$\frac{pd}{d+pm} < q \le p . \quad (2.13)$$

Then

$$\|f\|_{L_p} \le c\left(\|f\|_{L_q} + \|(-\Delta)^{m/2} f\|_{L_q}\right) \tag{2.14}$$

For $d \ge 3$, $q = 2$ *and* $pd = 2(d+pm)$, $m = 1$,

$$\|f\|_{L_{2d/(d-2)}} \le c\,\|\nabla f\|_{L_2} \quad . \tag{2.15}$$

Remark. The dimension d which satisfies $pd = q(d+mp)$ is called the critical dimension d_c for inequality (2.14). We can assign a dimension α to each term in (2.14), as the homogeneity in λ of the term for the function $f_\lambda(x) = f(\lambda x)$. For example define

$$\alpha(k,p) = \dim \|(-\Delta)^{k/2}\cdot\|_{L_p} = k - d/p \quad .$$

Clearly the dimensions of the three terms in (2.14) must satisfy

$$\alpha(0,q) \le \alpha(0,p) \le \alpha(m,q) \quad , \tag{2.16}$$

or else $\lambda \to 0$ or $\lambda \to \infty$ would provide a contradiction. The case $d = d_c$ corresponds to equality $\alpha(0,p) = \alpha(m,q)$. Under this circumstance, it is reasonable on dimensional grounds that the L_q norm of f may not be necessary for the inequality (2.14). For $q = 2$, $m = 1$, we prove this, i.e. (2.15). We do not prove other cases of (2.14) with $d = d_c$.

For $d < d_c$, we establish (2.14) under two circumstances: either with $q = 2$ or with $q \ge 2$ and $m = 1$. These cases, along with (2.15) are the only Sobolev inequalities needed for earlier chapters. Proofs for the general case of (2.14) can be found in [1,4,7,9,10].

Proof. First consider (2.14) with $q = 2$. As in the proof of Proposition 2.3, we have

$$\|f\|_{L_p}^p \le \|\tilde{f}^{p'}(1+k^2)^{\alpha}\|_{L_r}^{1/p'} \left(\int (1+k^2)^{-\alpha r'}\, dk\right)^{1/p'r'} \quad . \tag{2.17}$$

Choose r, α so that

$$r\alpha = m \qquad \text{and} \qquad rp' = 2 \quad . \tag{2.18}$$

The inequality $d < d_c$ means

$$2\alpha r' = 2\frac{mp}{p-2} > d \quad . \tag{2.19}$$

Thus (without keeping track of the d- or p-dependence of the constants)

$$\|f\|_{L_p} \le \text{const}\, \|\tilde{f}^{p'}(1+k^2)^{\alpha}\|_{L_r}^{1/p'} \quad . \tag{2.20}$$

Since $rp' = 2$, $r\alpha = m$,

$$\|f\|_{L_p} \le \text{const}\left(\|f\|_{L_2} + \|(-\Delta)^{m/2} f\|_{L_2}\right), \tag{2.21}$$

which is (2.14) for $q = 2$.

To replace L_2 in (2.21) by L_q, $q \ge 2$, consider $m = 1$. Note by Fourier transformation,

$$\|(-\Delta)^{1/2} f\|_{L_2} = \|\nabla f\|_{L_2} \quad .$$

Substitute $f = h^{q/2}$. By Hölder's inequality,

$$\|\nabla f\|_{L_2} = \frac{q}{2}\|h^{(q-2)/2}\nabla h\|_{L_2} \le \frac{q}{2}\|\nabla h\|_{L_q}\, \|h\|_{L_q}^{\frac{q-2}{2}} \quad .$$

Hence (2.21) can be written

$$\|h\|_{L_{qp/2}}^{q/2} \le c\left(\|h\|_{L_q} + \|\nabla h\|_{L_q}\right)\|h\|_{L_q}^{(q-2)/2}$$

$$\leq c\left(\|h\|_{L_q} + \|\nabla h\|_{L_q}\right)^{q/2} .$$

The renaming of $qp/2 \geq q$ as p then yields (2.14) with $m = 1$, and with $pd < q(d+p)$, as desired.

The method above is not applicable for $d = d_c$. To prove (2.15), choose $f \in C_0^\infty$. To be concrete, we give the proof for $d = 3$, though the proof for $d \geq 3$ involves minor change. Let (x,y,z) denote the coordinates in $\mathbb{R}^3$, and let $v = f^4$. Then

$$v(x,y,z) = \int_{-\infty}^{x} \partial_1 v(\xi,y,z)\, d\xi \leq \int_{-\infty}^{\infty} |(\partial_1 v)(\xi,y,z)|\, d\xi .$$

Then

$$|v|^{3/2} \leq \left(\int_{-\infty}^{\infty} |(\partial_1 v)(\xi,y,z)|\, d\xi \int_{-\infty}^{\infty} |(\partial_2 v)(x,\xi,z)|\, d\xi \int_{-\infty}^{\infty} |(\partial_3 v)(x,y,\xi)|\, d\xi\right)^{\frac{1}{2}}$$

Integrate x and then y and use Hölder's inequality to obtain

$$\int |v(x,y,z)|^{3/2}\, dxdy \leq \left(\int |(\partial_1 v)(\xi,y,z)|\, dyd\xi\right)^{1/2} \left(\int |(\partial_2 v)(x,\xi,z)|\, dxd\xi\right)^{1/2}$$
$$\left(\int |\partial_z v(x,y,\xi)|\, dxdyd\xi\right)^{1/2} .$$

Last, integrate z to obtain

$$\int |v|^{3/2}\, dxdydz \leq \left(\int |\nabla v|\, dxdydz\right)^{3/2} .$$

Since $v = f^4$, $\nabla v = 4f^3 \nabla f$, and one application of Hölder's inequality yields (2.15).

In §1 we defined Sobolev spaces L_p^k and norms $\|\cdot\|_{k,p}$, etc. A consequence of the inequalities (2.14-2.15) are inequalities between various of these norms. These inequalities can also be interpreted as proving imbeddings of L_p^k in $L_{p'}^{k'}$ for certain values of k, p, k', p'. For example, if there is a constant $c < \infty$ such that for all $f \in L_p^k$,

$$\|f\|_{k',p'} \le c\,\|f\|_{k,p} \quad , \tag{2.22}$$

then

$$L_p^k \hookrightarrow L_{p'}^{k'} \quad , \tag{2.23}$$

where $\hookrightarrow$ denotes a continuous inclusion map. We state the relations which follow by (2.14).

PROPOSITION 2.5. (Sobolev Imbeddings) *There exist constants* $c = c(m,p,q,d)$ *such that for all* $f \in L_q^{j+m}$ *and for* $pd/(d+pm) < q \le p$,

$$\|f\|_{j,p} \le c\,\|f\|_{j+m,q} \quad , \tag{2.24}$$

and

$$L_q^{j+m} \hookrightarrow L_p^j \quad . \tag{2.25}$$

The imbeddings are more useful when they are compact or Hilbert-Schmidt, as well as continuous. This requires restriction to a bounded subset V. We state without proof:

PROPOSITION 2.6. *Let* V *be an open, bounded domain with smooth boundary, and let*

$$\frac{pd}{d+pm} < q \le p \quad , \qquad 0 < m \quad .$$

Then the space $L_q^{j+m}(V)$ *is compactly imbedded in* $L_p^j(V)$.

Useful consequences of the Sobolev inequalities (2.14), (2.25) are the boundedness, differentiability, pointwise continuity of certain L_p^k functions. We summarize some local properties:

COROLLARY 2.7. *There exist constants* $c = c(d,m,p)$ *such that the following statements hold:*

(a) *Let* $mq > d + jq$. *Then for all* $f \in L_q^m$,

$$\max_{0 \le |i| \le j} \sup_{x \in \mathbb{R}^d} |\nabla^{(i)} f| \le c \, \|f\|_{m,q} \quad . \tag{2.26}$$

(b) *Let* $p > d$, $0 < \alpha \equiv 1 - d/p$. *Then every* $f \in L_p^1$ *is Hölder continuous with exponent* α*; i.e.*

$$|f(x) - f(y)| \le c|x - y|^{\alpha} \, \|\nabla f\|_{L_p} \quad . \tag{2.27}$$

Proof. The inequality (2.26) is obtained from (2.24) by letting $p = \infty$, $m \to m - j$. The proof of (2.27) proceeds as follows: Integrate the inequality

$$|f(x) - f(y)| \le |f(x) - f(z)| + |f(z) - f(x)|$$

over a ball B in the z variable with radius $|x - y|/4$ centered at $(x + y)/2$. Thus with $|B|$ the volume of B,

$$|B| \; |f(x) - f(y)| \le \int_B \{|f(x) - f(z)| + |f(z) - f(y)|\} \, dz \quad . \tag{2.28}$$

Let $x(t) = tx + (1-t)z$ parametrize the straight line from x to z. Then

$$|f(x) - f(z)| \le \int_0^1 \left| \frac{d}{dt} f(x(t)) \right| dt \quad . \tag{2.29}$$

Inserting (2.2.9) and a similar representation for $|f(z) - f(y)|$ into (2.28) yields, by Hölder's inequality in the z integration,

$$|f(x) - f(y)| \le 2|x - y| \; |B|^{-1+1/p'} \, \|\nabla f\|_{L_p} \quad .$$

Since $-1+1/p' = -1/p$, and $|B| = \Omega(|x-y|/4)^d$,

$$|f(x) - f(y)| \le \text{const}\,|x-y|^{1-d/p}\,\|\nabla f\|_{L_p} \quad .$$

Thus with $0<\alpha = 1-d/p$, the inequality (2.27) holds.

We end this section with some elementary estimates on convolutions. Define for $x \in \mathbb{R}^d$,

$$\mu(x) = (1+|x|^2)^{1/2} \sim \begin{cases} 1 \ , & |x| \to 0 \\ |x| \ , & |x| \to \infty \end{cases} \quad .$$

PROPOSITION 2.8. *Let* $\alpha + \beta > d$,

$$\nu = \min\{\alpha, \beta, \alpha+\beta-d\} \quad . \tag{2.30}$$

Then there is a constant $M<\infty$ *such that*

$$\int_{\mathbb{R}^d} \mu(x-y)^{-\alpha}\mu(y)^{-\beta}\,dy \le M \begin{cases} \mu(x)^{-\nu}\ell n\ \mu(x), & \textit{if}\ \max(\alpha,\beta) = d \\ \mu(x)^{-\nu} & , \textit{otherwise} \end{cases} \tag{2.31}$$

<u>Proof</u>. It is sufficient to consider $0<\alpha\le\beta$, for if $\alpha<0$, the proposition follows from the fact that $\nu = \alpha$ and

$$\mu(x-y)^{-\alpha} \le 2^{-\alpha}\mu(x)^{-\alpha}\mu(y)^{-\alpha} \quad . \tag{2.32}$$

We break the region of integration into two subsets $<$ and $>$ defined by

$$\begin{aligned} < \ &\equiv \{y\colon |x-y| < |x|/2\} \\ > \ &\equiv \{y\colon |x-y| > |x|/2\} \end{aligned} \quad . \tag{2.33}$$

On the set $<$, we have the upper and lower bounds

$$|x|/2 \le |y| \le 3|x|/2\,, \qquad \text{on} \quad <, \tag{2.34}$$

since

$$|y| = |y-x+x| \le |y-x| + |x| \le 3|x|/2 \,,$$

and

$$|y| = |y-x+x| \ge |x| - |y-x| \ge |x|/2 \,.$$

Thus in region $<$, $\mu(y) \ge \text{const}\ \mu(x)$; hence we infer

$$\int_< \mu(x-y)^{-\alpha}\mu(y)^{-\beta}\,dy \le \text{const}\ \mu(x)^{-\beta}\int_< \mu(x-y)^{-\alpha}\,dy$$

$$\le \text{const}\ \mu(x)^{-\beta}\begin{cases} 1\,, & \alpha > d \\ \ell n\ \mu(x)\,, & \alpha = d\,, \\ \mu(x)^{d-\alpha}\,, & \alpha < d \end{cases} \tag{2.35}$$

the desired bound for region $<$.

In region $>$, we have $\mu(x-y) \ge \text{const}\ \mu(y)$, when $|y| > 2|x|$. Then

$$\int_> \mu(x-y)^{-\alpha}\mu(y)^{-\beta}\,dy \le \text{const}\int_> \mu(y)^{-\alpha-\beta}\,dy$$

$$\le \text{const}\ \mu(x)^{-(\alpha+\beta-d)} \,.$$

If $|y| < 2|x|$, use $\mu(x-y) \ge \text{const}\ \mu(x)$ to complete the proof.

PROPOSITION 2.9. *Let* $\alpha+\beta < d$. *Then with* ν *defined in* (2.30),

$$\int_{|y|\le R} \mu(x-y)^{-\alpha}\mu(y)^{-\beta}\,dy \le \text{const}\ \mu(R)^{d-\alpha-\beta}\begin{cases} 1 & ,\ \alpha \ge 0 \\ \mu(x)^{|\alpha|} & ,\ \alpha < 0 \end{cases} \,. \tag{2.36}$$

If $\alpha+\beta = d$, $0 < \alpha,\beta$, *then*

$$\int_{|y|\le R} \mu(x-y)^{-\alpha}\mu(y)^{-\beta}\,dy \le \text{const } \ell n\, \mu(R) \quad . \tag{2.37}$$

Proof. First consider (2.37). By differentiation, we find that (2.37) attains its maximum at $|x| = 0$. But

$$\int_{|y|<R} \mu(y)^{-\alpha-\beta}\,dy = \int_{|y|<R} \mu(y)^{-d}\,dy$$

$$\le \text{const } \ell n\, \mu(R) \quad ,$$

as desired. To establish (2.36), note again by differentiation that for $\alpha \ge 0$, $|x| = 0$ is the maximum of the integral. Thus

$$\int_{|y|\le R} \mu(y)^{-\alpha-\beta}\,dy \le \text{const } \mu(R)^{d-\alpha-\beta}$$

yields the desired bound. If $\alpha < 0$, use $\mu(x-y)^{-\alpha} \le \text{const } \mu(x)^{-\alpha}\mu(y)^{-\alpha}$ and proceed as above.

VI.3. THE MAXIMUM PRINCIPLE

Let $\mathscr{L}$ be a second order operator of the form

$$\mathscr{L} = a^{ij}\nabla_i\nabla_j + \beta^j\nabla_j - c \tag{3.1}$$

on an open set $V \subset \mathbb{R}^d$. If the matrix a^{ij} is positive definite and $c(x) \ge 0$ then a superposition u ($\mathscr{L}u \le 0$) of (3.1) cannot achieve an infimum $u_0 \le 0$ in V. In outline, if u achieved a nonpositive infimum at $x \in V$ then $\beta^j(x)(\nabla_j u)(x) = 0$ and $a^{ij}\nabla_i\nabla_j u \ge 0$, which, we show below, is in contradiction with $\mathscr{L}u \le 0$.

For the remainder of §3 assume that the operator $\mathscr{L}$ is uniformly elliptic. That is, there exist constants $0 < \lambda_1 \le \lambda_2 < \infty$ such that for all $x \in V$ and unit vectors ξ,

$$0 < \lambda_1 \le a^{ij}\xi_i\xi_j \le \lambda_2 < \infty \tag{3.2}$$

It is often the case that u is strictly positive in the sense that either $u \equiv 0$ in V or $u > 0$ in V. Conditions on the operator $\mathscr{L}$ which imply this strong form of the maximum principle, as well as the following proof, are due to E. Hopf.

PROPOSITION 3.3. (The Maximum Principle) *Suppose that* $V \subseteq \mathbb{R}^d$ *is an open region with smooth (or empty) boundary* $\partial\bar{V}$. *Let* $u \in C^2(V)$, a^{ij}, β^j, $c \in C^0(\bar{V})$ *and* a^{ij} *uniformly elliptic. Let* $\mathscr{L}$ *be given by* (3.1) *with* $c \ge 0$ *and assume that*

$$\mathscr{L}u \le 0 \quad .$$

Suppose that $x_0 \in V$ *and* $u(x_0) = \inf_{x \in V} u(x)$. *Then either* $0 < u(x_0)$ *or else* $u(x_0) \equiv u(x)$. *In other words, an interior minimum of* u *must be strictly positive, or else* u *must be a constant.*

Proof. The idea is to prove that if $u(x_0) \le 0$, then $\Omega \equiv \{x \in V: u(x) > u(x_0)\}$ is both open and closed in V. That Ω is open is clear. Suppose the inequality $\mathscr{L}u < 0$. Then Taylor's expansion of $u(x)$ about x_0 proves that Ω is also closed. Hence $\Omega = \{\emptyset\}$ and $u(x) \equiv u(x_0)$.

To reduce $\mathscr{L}u \le 0$ to the case already proved, we add to u a small perturbation. Assume $u(x) \not\equiv u(x_0)$. Because Ω is open there exists a point x_2 at the center of an open ball $B \subset \Omega$ such that exactly one boundary point x_1 of B satisfies $u(x_1) = u(x_0)$. There is also a ball $B' \subset V$, centered at x_1, and with volume $|B'| < |B|$. Let $S_i = \bar{B} \cap \partial\bar{B}$, $S_e = \partial\bar{B}' \setminus S_i$. Then $u(x) - u(x_0) \ge \varepsilon > 0$ on S_i and $u(x) \ge u(x_0)$ on S_e.

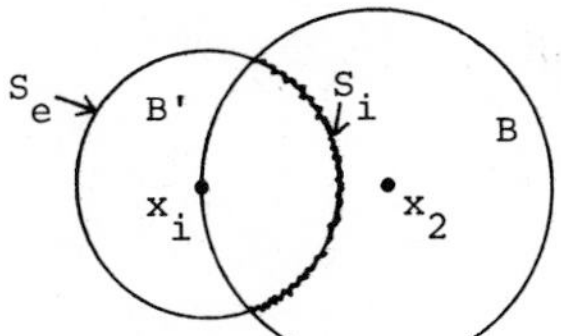

Define a comparison function

$$h(x) = \exp(-\alpha|x-x_2|^2) - \exp(-\alpha|x_1-x_2|^2) \quad .$$

For α sufficiently large,

$$h \restriction S_e < 0, \qquad \mathscr{L} h \restriction B' > 0 \quad .$$

Thus for $0 < \delta$, sufficiently small,

$$s \equiv u - \delta h \restriction S_i \cup S_e > u(x_0) \quad ,$$

$$\mathscr{L} s \restriction B' < 0 \quad \text{and} \quad s(x_1) = u(x_0) \quad .$$

By the special case proved above and applied to s on the domain B', we conclude $s(x) > 0$, because $s(x) \equiv u(x_0)$ is excluded by construction. Since $h(x_1) = 0$, this means $u(x_0) = u(x_1) > 0$.

Next we state a different form of the maximum principle for a related equation.

PROPOSITION 3.2. (Weak Maximum Principle) *Let* $V \subseteq \mathbb{R}^d$ *be an open set with smooth (or empty) boundary* $\partial\bar{V}$. *Suppose that* $\nabla u \in L_2(V)$, $u \in L_p(V)$ *for some* $2 \le p \le 2d/(d-2)$ *and that for all*

$$0 \le \xi \in L^1_{2(0)}(V) \quad .$$

$$\int_V \left\{(\nabla_i \xi - \beta_i \xi) a^{ij} (\nabla_j u - \beta_j u) + c\xi u\right\} \ge 0 \quad . \tag{3.3}$$

Assume that $0 \le c(x)$, $a^{ij}(x)$ *is uniformly elliptic and*

$$u_- = \min(u,0) \in L^1_{2(0)}(V) \quad . \tag{3.4}$$

Then

$$0 \le u \quad \textit{in } V \quad . \tag{3.5}$$

Because we allow unbounded V, it is convenient to introduce a cut-off function.

$$b_R(x) = b(x/R) \quad , \tag{3.6a}$$

where

$$b(x) \in C_0^\infty(|x| < 3/2) \quad , \tag{3.6b}$$

$$0 \le b(x) \le 1 \quad , \tag{3.6c}$$

$$b(x) = 1 \, , \qquad |x| \le 5/4 \quad . \tag{3.6d}$$

LEMMA 3.3. *Let* $1 \le q \le \infty$, $0 \le |s|$. *Then*

$$\|\nabla^{(s)} b_R\|_{L_q} = R^{d/q-|s|} \|\nabla^{(s)} b\|_{L_q} \quad . \tag{3.7}$$

Proof. The lemma follows from the scaling relation $(\nabla^{(s)} b_R)(x) = R^{-|s|}(\nabla^{(s)} b)(x/R)$.

Proof of Proposition 3.1. Let $u_- = \min(u,0)$. We show that $\Omega = \operatorname{supp} u_- = \emptyset$. By (3.4),

$$\xi = -b_R^2 u_- \in L^1_{2(0)}(V) \quad .$$

Substituting ξ into (3.3) we obtain

$$- \int_\Omega b_R^2 \left\{ (\nabla_i u - \beta_i u) a^{ij} (\nabla_i u - \beta_i u) + cu^2 \right\} \ge$$
$$+ 2 \int_\Omega b_R u \nabla_i b_R a^{ij} (\nabla_i u - \beta_i u) \quad . \tag{3.8}$$

Applying the triangle inequality to the right hand side of (3.8) gives

$$\int_\Omega b_R^2 \left\{ \frac{1}{2} (\nabla_i u - \beta_i u) a^{ij} (\nabla_j u - \beta_j u) + cu^2 \right\}$$
$$\le 4 \int_\Omega |u|^2 (\nabla_i b_R a^{ij} \nabla_j b_R) \quad . \tag{3.9}$$

Now use the fact that $(1-b_{R/2})\nabla b_R = \nabla b_R$, and Hölder's inequality to obtain

$$\int_\Omega b_R^2 \left\{\frac{\lambda_1}{2} |\nabla_i u-\beta_i u|^2 + c^2u^2\right\} \le 2\lambda_2 \,||(1-b_{R/2})u||^2_{L_p} \,||\nabla b_R||^2_{L_{2p/(p-2)}} ,$$

$$\le 2\lambda_2 (R^{d(p-2)/p-2}) \,||(1-b_{R/2})u||_{L_p}$$

$$||\Delta b||^2_{L_{2p/(p-2)}} . \quad (3.10)$$

The last step in (3.10) uses Lemma 3.2. Because $2 \ge d(p-2)/p$, the right hand of (3.10) is a strictly decreasing function of R. We take lim inf as $R \to \infty$ to conclude that $\Omega = \emptyset$. Thus $u \ge 0$ as claimed.

VI.4. GREEN'S FUNCTIONS AND POTENTIALS

Solutions to Laplace's equation on $\mathbb{R}^3$,

$$\Delta u = v \quad (4.1)$$

are formally given by

$$u(x) = -(Gv)(x) = -\int G(x-y)v(y)\, dy \quad . \quad (4.2)$$

For $d \ge 3$, the Green's function $G(x-y)$ is given by

$$0 < G(x-y) = [(d-2)\Omega(d-1)]^{-1}|x-y|^{-d+2} ,$$

where $\Omega(d-1)$ is the volume of the $(d-1)$ sphere. We give some simple conditions on v to ensure integrals such as (4.2) exist and are well behaved. Note that for any $\alpha < d/(d-2)$, $|G|^\alpha$ is locally integrable. Also for $\alpha > d/(d-2)$, $|G|^\alpha$ is integrable at infinity. The dual exponent to $d/(d-2)$ is $d/2$, cf. §2.

PROPOSITION 4.1. *Let* $d \geq 3$. *Assume* $v \in L_p \cap L_q$ *where* $p < d/2 < q$. *Then* $(Gv)(x)$ *exists and is continuous. If* $q < \infty$, *then* Gv *converges to zero uniformly,*

$$\lim_{R \to \infty} \sup_{|x|=R} |(Gv)(x)| = 0 \quad . \tag{4.3}$$

If $q = \infty$ *and in addition*

$$\lim_{R \to \infty} \sup_{|x|=R} |v(x)| = 0 \quad , \tag{4.4}$$

then (4.3) *holds. Furthermore if* $v \in L_p^k, L_q^k$ *for* $k > 0$, *then* Gv *is* C^{k+1}.

Proof. Divide the integration in (4.2) into two parts, according to whether $|x-y| < |x|/2$. Let

$$\begin{aligned} < \; &\equiv \{y: |x-y| < |x|/2\} \quad , \\ > \; &\equiv \{y: |x-y| > |x|/2\} \quad . \end{aligned} \tag{4.5}$$

On $<$ we have lower and upper bounds on $|y|$, namely

$$|x|/2 \leq |y| \leq 3|x|/2 \quad , \tag{4.6}$$

since

$$|y| = |x + y - x| \leq |x| + |y-x| \leq 3|x|/2 \tag{4.7}$$

and also

$$|y| = |x + y - x| \geq |x| - |y-x| \geq |x|/2 \quad . \tag{4.8}$$

Also note for p', q' the dual indices to p,q, that

$$q' < d/(d-2) < p' \quad . \tag{4.9}$$

Also if $v \in L_r$, $r < \infty$, then

$$\lim_{R\to\infty} \int_{|y|\geq R} |v(y)|^r \, dy = 0 \quad . \tag{4.10}$$

Now estimate Gv.

$$|(Gv)(x)| \leq \left(\int_< G(x-y)^{q'} \, dy\right)^{1/q'} \left(\int_{|y|>|x|/2} |v(y)|^q \, dy\right)^{1/q}$$

$$+\left(\int_> G(x-y)^{p'} \, dy\right)^{1/p'} \left(\int |v(y)|^p \, dy\right)^{1/p}$$

$$\leq \text{const} \left(\|v\|_{L_q} + \|v\|_{L_p}\right) \quad . \tag{4.11}$$

The constant in (4.11) depends on $|x|$. In fact, the estimate (4.11) shows that the integral (4.2) is absolutely convergent. To prove continuity of Gv, apply the same argument to $v - v_a$ where $v_a(x) = v(x-a)$ is the translate of v. Since translation is continuous in L_q for all $q \leq \infty$, cf. Proposition 1.11, it follows that $(Gv)(x)$ is continuous for $q < \infty$. With $q = \infty$, use the continuity of $\int_< G(x-y)\, dy$ in x. Hence we have established the existence and continuity of Gv.

To establish the asymptotic behavior of Gv, we replace the regions "<" and ">" by regions

$$<\rho \equiv \{y: |x-y| < \rho\} \quad , \text{ etc.} \tag{4.12}$$

We choose ρ large, and $|x| > R + \rho$ so that if $y \in <\rho$,

$$|y| = |x + y - x| \geq R + \rho - \rho = R \quad . \tag{4.13}$$

Then as in (4.11), but for $|x| > R$, obtain for constants c_1, c_2

$$|(Gv)(x)| \le c_1 \rho^{(d/q'-d+2)} \left(\int_{R<|y|} |v(y)|^q \, dy \right)^{1/q}$$

$$+ c_2 \rho^{-(d-2-d/p')} \|v\|_{L_p} \quad . \tag{4.14}$$

Condition (4.9) shows $d/q' - d + 2 > 0$, and $d - 2 - d/p' > 0$. Let

$$c_2 \rho^{-(d-2-d/p')} \|v\|_{L_p} = \varepsilon \quad .$$

Then

$$|(Gv)(x)| \le c_1 \left(\frac{c_2 \|v\|_{L_p}}{\varepsilon} \right)^{\left(\frac{d/q'-d+2}{d-2-d/p'}\right)} \left(\int_{R<|y|} |v(y)|^q \, dy \right)^{1/q} + \varepsilon \ .$$

Choose R sufficiently large so

$$|(Gv)(x)| < 2\varepsilon \quad .$$

This proves the uniform decay of $|(Gv)(x)|$, when $q < \infty$. If $q = \infty$, we replace the $R < |y|$ integral in (4.14) by

$$\sup_{|y| \ge R} |v(y)| \ .$$

Using (4.4), we obtain $|(Gv)(x)| < \varepsilon$.

Finally, the smoothness of Gv follows from the local integrability of $G(x-y)$, $(\nabla G)(x-y)$ and the decay of v assumed. The smoothness follows by integration by parts and similar estimates to the above.

COROLLARY 4.2. *Let* v *be as in Proposition* 4.1. *Then* $-Gv$ *is the unique solution to* $\Delta u = v$ *which vanishes at* ∞.

Proof. Let u be another solution to (4.1) which vanishes at infinity. Then $h \equiv u - Gv$ satisfies $\Delta h = 0$, $h \to 0$ at ∞. By the maximum principle, h achieves its maximum and minimum values at ∞. Thus $h \equiv 0$.

Next we make a somewhat stronger assumption concerning the uniformity of decay of v.

PROPOSITION 4.3. *Let* $d \geq 3$. *Let* $0 < r < d-2$ *and assume*

$$|v(x)| \leq (1+|x|)^{-(2+r)} \quad . \tag{4.15}$$

Then Gv *exists. If* $\varepsilon > 0$, *there exists* $c = c(\varepsilon) < \infty$ *such that*

$$|(Gv)(x)| \leq c(1+|x|)^{-r+\varepsilon} \quad . \tag{4.16}$$

If $v \in L_{d/(r+2)}$, *then* (4.16) *holds with* $\varepsilon = 0$. *If* (4.15) *holds with* $r > d-2$, *then*

$$|(Gv)(x)| \leq c(1+|x|)^{-d+2} \quad . \tag{4.17}$$

Proof. We first verify the hypotheses of Proposition 10.1, to ensure the existence of Gv. In fact $(2+r)d/2 > d$, so the appropriate p, q exist with $p'(d-2)/d \geq 1+\varepsilon > 1$. As in (4.11), there are constants c_1, $c_2 < \infty$ such that

$$|(Gv)(x)| \leq c_1(1+|x|)^{-r} + c_2 \|v\|_{L_p} |x|^{-d+2+d/p'} \quad .$$

In the first term we used $-d+2+d/q'+d/q-r-2 = -r$. In the second term we used

$$\left(\int_{|x-y|>|x|/2} |x-y|^{-(d-2)p'} dy\right)^{1/p'}$$

$$\leq \left(\Omega(d-1)((d-2)p'-d)^{-1}\right)^{1/p'} (|x|/2)^{-(d-2)+d/p'}$$

$$\leq \text{const}|x|^{-d+2+d/p'} \quad .$$

If $r > (d-2)$, take $p = 1$, $p' = \infty$, to obtain (4.1.7). For $r < (d-2)$, note $p(2+r) > d$. Thus for $0 < \varepsilon$, but close to zero $-d + 2 + d/p' = -r + \varepsilon$. If $v \in L_{d/(2+r)}$, we take $p = d/(2+r)$. Then $-d + 2 + d/p' = -r$. This gives (4.16) for $|x| \geq 1$. Since Gv is continuous, by Proposition 10.1, the bounds (4.16-17) hold for all x, as desired.

When the *a priori* estimates on v are insufficient to apply Proposition 4.2, we differentiate Gv to obtain bounds on the derivative. Let

$$H(x-y) = \nabla_x G(x-y) \tag{4.18}$$
$$(Hv)(x) = \int H(x-y)v(y)\, dy \quad .$$

PROPOSITION 4.4. *Let* $d \geq 3$. *Let* $0 < r < d-1$. *Suppose*

$$|v| \leq (1+|x|)^{-(1+r)} \quad . \tag{4.19}$$

Then $(Hv)(x)$ *is defined by* (4.18). *Also given* $\varepsilon > 0$, *there exists* $c < \infty$ *such that*

$$|(Hv)(x)| \leq c(1+|x|)^{-r+\varepsilon} \quad . \tag{4.20}$$

If $v \in L_{d/(1+r)}$, *then* (4.20) *holds with* $\varepsilon = 0$.

Proof. The derivative of the Green's function satisfies

$$|\nabla G(x-y)| \leq \text{const} \frac{1}{|x-y|^{d-1}} \quad .$$

Thus

$$\left| \int (H)(x-y)v(y)\, dy \right| \leq \text{const} \int \frac{|v(y)|}{|x-y|^{d-1}}\, dy \quad . \tag{4.21}$$

The estimate (4.18) ensures the convergence of (4.20). Using the notation of the proof of Proposition 4.1,

$$|(H)v(x)| \le \text{const}(1+|x|)^{-r} + \text{const}\int_{>} \frac{|v(y)|}{|x-y|^{d-1}}\, dy$$

$$\le \text{const}(1+|x|)^{-r} + \text{const}|x|^{-(d-1)+d/p'}$$

$$\le \text{const}(1+|x|)^{-r+\varepsilon}$$

since $v \in L_p$, $p = d/(1+r) + \varepsilon$. As before, if $v \in L_{d/(1+r)}$, we obtain the $\varepsilon = 0$ estimate.

COROLLARY 4.5. *Let* v *satisfy the hypotheses of the proposition. Then* $-Hv$ *is the unique solution which vanishes at infinity to the equation*

$$\Delta u = \nabla v \quad . \tag{4.22}$$

Proof. Let $b_R(x) = b(x/R)$ with $b \in C_0^\infty$, $b \equiv 1$ for $|x| \le 1$, and $0 \le b \le 1$. Then

$$(Hv)(x) = \lim_{R\to\infty} \int H(x-y)v(y)b_R(y)\, dy \quad . \tag{4.23}$$

In fact the dominated convergence theorem and the bound of the proposition allows exchange of the limit and integration. Since $H(x-y) = \nabla_x G(x-y) = -\nabla_y G(x-y)$ and b_R has compact support, we can integrate by parts to obtain

$$(Hv)(x) = \lim_{R\to\infty} \left[\int G(x-y)\nabla v(y) b_R(y)\, dy + \int G(x-y)v(y)\nabla b_R(y)\, dy \right] . \tag{4.24}$$

Note $\nabla b_R(y) = R^{-1}(\nabla b)(y/R)$. But by (4.19)

$$\int_{|y|<\xi} G(x-y)|v(y)|dy \le \text{const}(1+|\xi|)^{1-r} . \tag{4.25}$$

Thus it follows that

$$\int |G(x-y)v(y)(\nabla b_R)(y)|\, dy \;\le\; R^{-1}\,\mathrm{const}(1+R)^{1-r}$$

$$\le\; \mathrm{const}(1+R)^{-r} \to 0 \quad .$$

Hence by (4.23-24),

$$(Hv)(x) \;=\; \lim_{R\to\infty} \int G(x-y)(\nabla v)(y)\, b_R(y)\, dy \quad . \qquad (4.26)$$

Since both (4.23) and (4.26) are independent of R as long as $|x| < \mathrm{const}\ R$, and

$$\Delta(Hv)(x) \;=\; \nabla v$$

solves the equation $\Delta u = \nabla v$ as claimed. Uniqueness follows as in Corollary 4.2.

Note that Proposition 4.3 yields at best $|x|^{-(d-2)}$ decay at ∞, for u satisfying $\Delta u = v$. Proposition 4.4 gives at most $|x|^{-(d-1)}$ decay for u, when $\Delta u = \nabla v$. Here we give conditions on v so that u, satisfying $\Delta u = v$, decays as $|x|^{-(d-1)}$. Essentially, we justify writing $v = \nabla w$ where w satisfies the conditions of Proposition 4.4.

PROPOSITION 4.6. *Let* $d \ge 3$. *Let* $v \in C^\infty$ *satisfy for some* $0 \le r < 1$,

(a) $|v(x)| \;\le\; (1+|x|)^{-(d+1)+r}$

(b) $|x|v \in L_{d/(d-r)}$

(c) $\int_{\mathbb{R}^d} v \;=\; 0 \quad . \qquad (4.27)$

Then Gv *exists, and there exists a constant* $c < \infty$ *such that*

$$|(Gv)(x)| \;\le\; c(1+|x|)^{-(d-1)+r} \quad . \qquad (4.28)$$

Also $u = -Gv$ *is the unique solution to* $\Delta u = v$ *which vanishes at infinity.*

Proof. Since $G = \text{const}|x-y|^{-d+2}$ and for some $\varepsilon > 0$, $|v| \le (1+|x|)^{-d-\varepsilon}$, the convolution Gv exists; furthermore $v \in L_1$. Thus for $x \ne 0$ we use (c) to write

$$(Gv)(x) = \int [G(x-y) - G(x)]v(y)\,dy \quad . \tag{4.29}$$

By the fundamental theorem of calculus, for $x \ne 0$,

$$\frac{1}{|x-y|^{d-2}} - \frac{1}{|x|^{d-2}} = \int_0^1 \frac{d}{dt}\left(|x-ty|^{-d+2}\right) dt$$

$$= (d-2)\int_0^1 \frac{(x-ty)\cdot y}{|x-ty|^d}\,dt \quad .$$

Therefore

$$|G(x-y) - G(x)| \le \text{const}|y| \int_0^1 \frac{dt}{|x-ty|^{d-1}} \quad . \tag{4.30}$$

Inserting (4.30) into (4.29) gives

$$|(Gv)(x)| \le \text{const} \int_{\mathbb{R}^d} \left(\int_0^1 \frac{|y||v(y)|}{|x-ty|^{d-1}}\,dt \right) dy \quad . \tag{4.31}$$

The integrand of (4.31) is positive so the order of integration can be interchanged. Break the y integration into two regions $<$ and $>$, as in (4.5), but with $|x-ty|$ replacing $|x-y|$. Thus by the methods of the previous proofs,

$$|(Gv)(x)| \le \text{const} \int_0^1 dt \left\{ t^{-r}|x|^{-(d-1)+r} + t^{-r}|x|^{-(d-1)+r}\,\|yv(y)\|_{L_{d/d-r}} \right\} . \tag{4.32}$$

In the estimate of the $>$ term we used $(d/(d-r))' = d/r$, and $(d-1)d/r > 1$. Thus $|x-ty|^{-d+1} \in L_{(d/d-r)'}$. Furthermore, for the $<$ term, we used $|ty| > |x|/2$ so by (a)

$$|yv(y)| \le \text{const}(1+|x|/t)^{-d+r}$$

$$= \text{const } t^{d-r}(t+|x|)^{-d+r}$$

$$\le \text{const } t^{d-r}|x|^{-d+r} \quad .$$

Also

$$\int_{<} \frac{1}{|x-ty|^{d-1}}\, dy \le \text{const } t^{-d}|x| \quad .$$

Since $r \le 1$, the t integration converges and for $|x| \ne 0$,

$$|(Gv)(x)| \le \text{const}|x|^{-(d-1)+r} \quad ,$$

as claimed. For $|x| \le 1$, Gv is bounded. Hence (4.28) holds for all x. The uniqueness result follows as before.

In applications of Proposition 4.6, we want to know *a priori* when (4.27c) holds. We now derive this from other properties.

PROPOSITION 4.7. *Let*

$$v \in L_1(\mathbb{R}^d) \cap L_\infty(\mathbb{R}^d) \tag{4.33}$$

and

$$Gv \in L_p(\mathbb{R}^d) \qquad \textit{some} \quad p < d/(d-2) \quad . \tag{4.34}$$

Then

$$\int_{\mathbb{R}^d} v = 0 \quad . \tag{4.35}$$

<u>Proof</u>. By (4.33) and Proposition 4.1, Gv exists and $u = -Gv$ satisfies

$$\Delta u = v \quad . \tag{4.36}$$

We multiply (4.36) by $b_R(x)$ and integrate by parts. Here $b_R(x)$ is as in the proof of Corollary 4.5. Thus

$$|\int b_R v| = |\int Gv\Delta b_R| \leq \|Gv\|_{L_p} \|\Delta G_R\|_{L_{p'}} . \tag{4.37}$$

Note

$$\|\Delta b_R\|_{L_{p'}} = \|\Delta b\|_{L_{p'}} R^{-(2-d/p')} . \tag{4.38}$$

But with $p' = p/(p-1)$, $d/p' = d(p-1)/p < 2$. Since $v \in L_1$, $\int v = \lim \int vb_R$. Hence by (4.37-38),

$$\left|\int_{\mathbb{R}^d} v\right| = \left|\lim_{R\to\infty} \int_{\mathbb{R}^d} b_R v\right| = 0 .$$

We conclude this section with a short exposition on the Green's function $C(x-y)$ for the operator $-\Delta + I$ on $\mathbb{R}^d$, $d \geq 2$. Let

$$(Cv)(x) = \int C(x-y)v(y)\, dy .$$

PROPOSITION 4.8. *For* $v \in L_2$, Cv *is* L_2. *The map*

$$C: L_2^k \to L_2^{k+2} \tag{4.39}$$

is bounded for any $k \geq 0$. *For* $v \in L_2$, Cv *is the unique solution to Laplace's equation* $(-\Delta + I)f = v$ *within the class of tempered distributions. If also* $v \in L_q$ *for some* $q > d/2$, *then*

$$\lim_{R\to\infty} \sup_{|x|=R} |(Cv)(x)| = 0 . \tag{4.40}$$

If $v \geq 0$, $v \not\equiv 0$, *then* $(Cv)(x) > 0$.

Proof. Since C is an isomorphism of the tempered distributions $\mathscr{S}'$, Cv is defined. Uniqueness of the solutions to Laplace's equation then follows from $f = 0$ being the unique tempered distribution solution to $(-\Delta + I)f = 0$. The boundedness of (4.39) follows by Fourier transformation. Since $C(x-y)$ is bounded by const $\exp(-|x-y|)$ for $|x-y| \geq 1$, it follows that with the notation of (4.12), that

$$\left|\int_{>\rho} C(x-y)f(y)\,dy\right| \leq O(1)e^{-\rho}\,\|f\|_{L_q} \quad .$$

Thus the decay of Cf depends on the decay of

$$\left|\int_{<\rho} C(x-y)f(y)\,dy\right|$$

which can be estimated using $C(x-y) \leq G(x-y)$ and the methods in (4.14). That $C(x-y) \leq G(x-y)$ follows from $0 \leq C(x-y)$ and

$$\frac{\partial}{\partial\lambda}(-\Delta+\lambda)^{-1}(x-y) = -\int C(\lambda;x-z)C(\lambda;z-y)\,dz \leq 0 \quad .$$

Finally, strict positivity, $0 < C(x-y)$, follows by direct computation of

$$C(x-y) = \left(\frac{1}{2\pi}\right)^d \int e^{-ikx}\,\frac{1}{k^2+1}\,dk \quad . \tag{4.41}$$

It also yields $(Cv)(x) > 0$ for $v \geq 0$, unless $v \equiv 0$.

VI.5. COVARIANT DERIVATIVES

Let $\mathbb{L}$ be an n-dimensional real vector space with positive inner product $(\ ,\ )$. We denote $E = \mathbb{R}^d \times \mathbb{L}$. A connection A on E is equivalent to a map, the covariant derivative D_A, from sections of E to sections of $T^* \otimes E$, where T^* is the cotangent bundle. For Φ a smooth cross-section of $T^* \otimes E$, namely $\Phi \in C^\infty(\mathbb{R}^d; T^* \otimes E)$,

$$D_A\Phi = d\Phi + A(\Phi) = \left(\frac{\partial}{\partial x^j}\Phi^a + A_j^{ab}\Phi^b\right) e_a dx^j \quad . \tag{5.1}$$

Here $\{e_a\}_{a=1}^n$ and $\{dx^j\}_{j=1}^d$ are orthonormal bases for $\mathbb{L}$ and T^*, respectively. The matrix A_j^{ab} for each $j = 1,\dots,d$ is defined by A; at each point it is an endomorphism of $\mathbb{L}$. We require that the inner product be compatible with D_A in the sense that for any pair $\sigma,\eta \in C^\infty(\mathbb{R}^d; E)$,

$$d(\sigma,\eta) = (D_A\sigma,\eta) + (\sigma,D_A\eta) \quad . \tag{5.2}$$

The inner products in Chapters I-V satisfy (5.2).

The inner product $(\, , \,)$ and the (flat) Riemannian metric induce an inner product on

$$\Lambda_p T^* \otimes E = \underbrace{T^* \wedge \ldots \wedge T^*}_{p} \otimes E, \quad p = 1,\ldots,d,$$

which is the bundle of p-forms with values in E.

This inner product on the fibers of $(\Lambda_p T^*) \otimes E$ will be denoted by $(\, , \,)$ also. The inner product is defined as follows: The Hodge dual map, denoted $*$, is an isomorphism from $\Lambda_p T^*$ to $\Lambda_{d-p}T^*$. If

$$\omega = dx^{i_1} \wedge \ldots \wedge dx^{i_p} \quad ,$$

then

$$*\omega = \frac{1}{p!} \varepsilon^{i_1 \cdots i_p j_1 \cdots j_{d-p}} dx^{j_1} \wedge \ldots \wedge dx^{j_{d-p}} \tag{5.3}$$

Here $1 \leq i_1,\ldots,i_p \leq d$ and $\varepsilon^{(i)}$ is the completely antisymmetric d-tensor with $\varepsilon^{1\,2\ldots d} = 1$. The Hodge dual extends by linearity to $\Lambda_p T^*$ and to an isomorphism $*\omega$: $(\Lambda_p T^*) \otimes E \to (\Lambda_{d-p}T^*) \otimes E$. A section ω of $\Lambda_p T^* \otimes E$ has the decomposition

$$\omega = \omega^a_{2_1,\ldots,i_p} dx^{i_1} \wedge \ldots \wedge dx^{i_p} \quad .$$

Then

$$(\omega,\eta) = *(\omega^a \wedge *\eta^a) \tag{5.4}$$

where $\{\omega^a\}_{a=1}^n$, $\{\eta^a\}_{a=1}^n$ are the components of ω,η with respect to an orthonormal frame for $\mathbb{L}$. On $(\Lambda_p T^*) \otimes E$,

$$** = (-1)^{p(d-p)} . \tag{5.5}$$

The covariant derivative D_A has two natural extensions to $\bigoplus_{p=0}^{d} C^\infty(\mathbb{R}^d; (\Lambda_p T^*) \otimes E)$. The first extension is a map from $C^\infty(\mathbb{R}^d; (\Lambda_{p+1} T^*) \otimes E)$ which is defined by

$$D_A \omega = (d\omega^a + A^{ab} \wedge \omega^b) e_a . \tag{5.6}$$

The formal adjoint of

$$D_A : C^\infty(\mathbb{R}^d; (\Lambda_p T^*) \otimes E) \to C^\infty(\mathbb{R}^d; (\Lambda_{p+1} T^*) \otimes E)$$

is

$$D_A^* = (-1)^{(p+1)(d-p)-1} *D_A* . \tag{5.7}$$

The covariant Laplace-Beltrami operator, Δ_A, associated with (D_A, D_A^*) is formed in the usual way. As an endomorphism of $C^\infty(\mathbb{R}^d; \Lambda_p T^* \otimes E)$,

$$\Delta_A = -(D_A^* D_A + D_A D_A^*) . \tag{5.8}$$

The second extension is a covariant derivative

$$\nabla_A : C^\infty(\mathbb{R}^d; (\Lambda_p T^*) \otimes E) \to C^\infty(\mathbb{R}^d; T^* \otimes ((\Lambda_p T^*) \otimes E))$$

defined by replacing E in (5.1) with $((\Lambda_p T^*) \otimes E)$. The Laplacian associated with ∇_A is the endomorphism of $C^\infty(\mathbb{R}^d; (\Lambda_p T^*) \otimes E)$ given by

$$(\nabla_A^2)^{ab} = (\nabla_A)_j^{ac} (\nabla_A)_j^{cb} \tag{5.9}$$

where

$$(\nabla_A)_j^{ab} = \delta^{ab} \frac{\partial}{\partial x^j} + A_j^{ab} .$$

For simplicity, we often write $(\nabla_A^2)^{ab} = \nabla_A^2$ and $(\nabla_A)_j^{ab} = \nabla_{Aj}$. Clearly, D_A and ∇_A agree on 0-forms as do Δ_A and ∇_A^2. Physicists generally write formulas with the derivative ∇_A.

The formula relating Δ_A and ∇_A^2 involves the curvature: The curvature of the connection A is the two-form with values in the vector space of endomorphisms of E,

$$F_A^{ab} = \frac{1}{2} F_{Ajk}^{ab}\, dx^j \wedge dx^k = \frac{1}{2} [(\nabla_A)_j, (\nabla_A)_k]\, dx^j \wedge dx^k . \tag{5.10}$$

PROPOSITION 5.1. *Let* A *be a smooth connection on* E *and* $\omega \in C^\infty(\mathbb{R}^d; (\Lambda_p T^*) \otimes E)$. *Then*

$$\Delta_A \omega = \nabla_A^2 \omega - pF_{kj_1}^{ab}\, \omega^b_{kj_2 \ldots j_p}\, dx^{j_1} \wedge \ldots \wedge dx^{j_p}\, e_a \quad . \tag{5.11}$$

Proof. The proposition follows from the following identities:

$$D_A * D_A * \omega = p(-1)^{(d-p)(p-1)} (\nabla_A)_{j_1}^{ac} (\nabla_A)_k^{cb} \omega^b_{kj_2 \ldots j_p}\, dx^{j_1} \wedge \ldots \wedge dx^{j_p}\, e_a$$

$$*D_A * D_A \omega = (-1)^{p(d-p-1)} \left(\nabla_A^2 \omega - p(\nabla_A)_k^{ac} (\nabla_A)_{j_1}^{ab} \omega^b_{kj_2 \ldots j_p}\, dx^{j_1} \wedge \ldots \wedge dx^{j_p} e_a \right) \tag{5.12}$$

The derivation of the above identities is an exercise in the manipulation of forms which we omit.

The curvature F_A measures the amount by which the components ∇_{Aj} fail to commute.

PROPOSITION 5.2. *Let* A *be a smooth connection on* E *and* $\omega \in C^\infty(\mathbb{R}^d; (\Lambda_p T^*) \otimes E)$. *Then*

(a) $D_A D_A \omega = F_A^{ab} \wedge \omega^b\, e_a$,

(b) $[(\nabla_A)_j, \nabla_A^2]\omega = (2F_{jk}^{ab}(\nabla_{A_k}\omega)^b + (\nabla_{A_k}F_{jk})^{ab}\omega^b)e_a$. (5.13)

Note that F_{ajk} as an $n \times n$ matrix at each point defines an endomorphism of $\mathbb{L}$ and hence is a section of the vector bundle of endomorphisms of $\mathbb{L}$. The covariant derivative extends naturally to this new vector bundle and it is in this way that $(\nabla_{A_k}F_{jk})^{ab}$ is defined:

$$(\nabla_{A_i}F_{jk})^{ab} = \nabla_i F_{jk}^{ab} + A_i^{ac}F_{jk}^{cb} + A_i^{cb}F_{jk}^{ac} . \tag{5.14}$$

<u>Proof</u>. As for statement (a),

$$D_A D_A \omega = \nabla_{A_j}\nabla_{A_k}\omega \wedge dx^j \wedge dx^k = \frac{1}{2}[\nabla_{A_j},\nabla_{A_k}]\omega \wedge dx^j \wedge dx^k .$$

Concerning statement (b):

$$\nabla_{A_j}\nabla_A^2 = \nabla_A^2\nabla_{A_j} + [\nabla_{A_j},\nabla_{A_k}]\nabla_{A_k} + \nabla_{A_k}[\nabla_{A_j},\nabla_{A_k}] .$$

As in Sections 1 and 2 we define L_p^k norms on sections of E. If $\Phi \in C_0^\infty(\mathbb{R}^d; E)$, set

$$\|\Phi\|_{k,p} = \left(\sum_{0\le|s|\le k}\int_{\mathbb{R}^d}(\nabla^{(s)}\Phi,\nabla^{(s)}\Phi)^{p/2}\right)^{1/p}, \quad k \ge 0,\ p > 0.$$

The space $L_p^k(\mathbb{R}^d; E)$ $(k \ge 0,\ p > 0)$ is the completion of $C_0^\infty(\mathbb{R}^d; E)$ in the norm $\|\cdot\|_{k,p}$. The spaces $L_2^k(\mathbb{R}^d; E)$ are Hilbert spaces with inner product

$$\langle\eta,\Phi\rangle_{H_k} = \sum_{0\le|s|\le k}\int_{\mathbb{R}^d}(\nabla^{(s)}\eta,\nabla^{(s)}\Phi) . \tag{5.15}$$

We remark that a section Φ is an element of $L_k^p(\mathbb{R}^d; E)$ iff its components with respect to some orthonormal basis e_a of $\mathbb{L}$ are in the

Banach spaces $L_k^p(\mathbb{R}^d)$ of functions introduced in §1. This implies that the Sobolev inequalities of §1,2 generalize word for word to the spaces $L_p^k(\mathbb{R}^d; E)$.

VI.6. KATO'S INEQUALITY

We continue with the notation of the previous section. The inner product $(\ ,\ )$ defines the absolute value map from $C^\infty(\mathbb{R}^d; E)$ to $C^0(\mathbb{R}^d)$ via

$$|\Phi|(x) = (\Phi(x),\Phi(x))^{1/2} \quad . \tag{6.1}$$

The function $|\Phi|$ is smooth on the complement of a set of measure zero, i.e., $|\Phi| \in C^\infty(\mathbb{R}^d \setminus \partial\overline{(\mathrm{supp}|\Phi|)})$. Therefore, $d|\Phi|$ can be defined almost everywhere. An L_p bound on the size of $d|\Phi|$ is given by

PROPOSITION 6.1. (Kato's Inequality) *Let* $\Phi \in C_0^\infty(\mathbb{R}^d, E)$, *and let* A *be a smooth connection on* E. *Then*

$$\langle D_A\Phi, D_A\Phi\rangle_{L_2} \equiv \int_{\mathbb{R}^d} (D_A\Phi, D_A\Phi) \geq \int_{\mathbb{R}^d} |d|\Phi||^2 \quad . \tag{6.2}$$

COROLLARY 6.2. *Let* A *be a smooth connection on* E. *Let* $d \geq 3$. *There exists a constant* $c(d) > 0$, *independent of* A, *such that for* $\Phi \in C_0^\infty(\mathbb{R}^d; E)$,

$$\|D_A\Phi\|_{L_2}^2 > c\|\Phi\|_{L_q}^2 \quad , \qquad q = 2d/(d-2) \quad . \tag{6.3}$$

COROLLARY 6.2. *Let* A *be a smooth connection on* E. *The operator*

$$-\nabla_A^2 : C_0^\infty(\mathbb{R}^d; E) \to C_0^\infty(\mathbb{R}^d; E)$$

satisfies

$$0 < \langle \Phi, -\nabla_A^2\Phi\rangle \tag{6.4}$$

unless $\Phi = 0$.

Proof of Corollaries 6.2 and 6.3. Corollary 6.2 follows from Proposition 2.8. Corollary 6.3 follows from (6.3), the identity $\|D_A\Phi\| = \|\nabla_A\Phi\|$ for zero forms, and an integration by parts.

Proof of Proposition 6.1. Define a smooth function $|\Phi|_\rho = ((\Phi,\Phi) + \rho^2)^{1/2}$, $\rho > 0$. Then

$$|d|\Phi|_\rho| = ||\Phi|^{-1}(\Phi, D_A\dot{\Phi})| \leq |D_A\Phi| \quad . \tag{6.5}$$

Because

$$\lim_{\rho\to 0} d|\Phi|_\rho = d|\Phi| \qquad \text{a.e.,} \tag{6.6}$$

$$\|d|\Phi|\|_{L_2} \leq \|D_A\Phi\|_{L_2} \quad . \tag{6.7}$$

PROPOSITION 6.4. *For* $\Phi \in C_0^\infty(\mathbb{R}^2; E)$, *the sequence* $\{d|\Phi|_{1/n}\}_{n=1}^\infty$ *converges strongly to* $d|\Phi|$ *in* L_2.

Proof. We prove that given $\varepsilon > 0$, there exists $N > 0$ such that for all $n > N$

$$\|d|\Phi| - d|\Phi|_{1/n}\|_{L_2} < \varepsilon$$

We use the identity

$$d|\Phi|_{1/n} = \frac{|\Phi|}{|\Phi|_{1/n}} d|\Phi| \quad . \tag{6.8}$$

Let $V_N = \{x \in \mathbb{R}^d : 0 < |\Phi(x)| < 1/\sqrt{N}\}$, $N \in \mathbb{Z}$. Choose N sufficiently large so that

$$\left[\int_{V_N} |D_A|^2\right]^{1/2} < \varepsilon/2 \tag{6.9}$$

and

$$\frac{1}{2N} \|D_A\Phi\|_{L_2} < \varepsilon/2 \quad . \tag{6.10}$$

For all $n > 1$,

$$\| d|\Phi|_{1/n} - d|\Phi| \|_{L_2} = \left\| \left(1 - \frac{|\Phi|}{|\Phi|_{1/n}}\right) |d|\Phi|| \right\|_{L_2} \quad ;$$

and for all $n > N$,

$$\| d|\Phi|_{1/n} - d|\Phi| \|_{L_2} \leq \left[\int_{V_N} |D_A\Phi|^2 \right]^{1/2} + \frac{1}{2N} \|D_A\Phi\|_{L_2} \quad . \tag{6.11}$$

Here we have used the inequality

$$1 - \frac{|\Phi|}{|\Phi|_{1/N}} \leq \frac{1}{2N} \qquad \text{for } |\Phi| > \sqrt{N} \quad . \tag{6.12}$$

Substituting (6.9) and (6.10) into (6.11) yields

$$\| d|\Phi|_{1/n} - d|\Phi| \|_{L_2} < \varepsilon \quad ,$$

as claimed.

VI.7. FUNCTIONALS

Suppose that a set of nonlinear equations are the variational equations of a functional defined on a named vector space. Then the problem becomes one of the calculus of variations. The existence of solutions to the equations is implied by the existence of critical points of the functional and vice-versa. For a linear space of finite dimension, a large number of techniques have been developed to find critical points. Many of these have been generalized to the case of infinite dimensional Banach spaces (see, e.g. [2,10]). The complication in dealing with an infinite dimensional space is controlled in most cases by requiring properties of the functional in question which in some sense ensure that the level sets are compact. This is made specific in §8.

Let X be a Banach space with norm $\|\cdot\|$.

DEFINITION 7.1. *A functional* A *on* X *is a map* $A: X \to \mathbb{R}$ *such that* $|A(v)| < \infty$ *for all* $v \in X$.

In discussing the continuity of a functional A, one must differentiate between two types of continuity, weak and strong, which agree if X is finite dimensional. Strong continuity is defined as follows. A functional A is strongly continuous if for every convergent sequence in X,

$$\|v_n - v\| \to 0 \qquad , \tag{7.1}$$

we can infer

$$\lim_{n\to\infty} |A(v) - A(v_n)| = 0 \qquad . \tag{7.2}$$

In order to define weak continuity, we introduce the dual space, X' to X.

The dual space X' is the space of bounded, linear functionals $\ell: X \to \mathbb{R}$. The space X' has a natural Banach space structure induced by X with the norm

$$\|\ell\|' = \sup_{\|v\|=1} \ell(v) \qquad . \tag{7.3}$$

The statement

$$\text{weak} \lim_n v_n = v, \qquad (v_n \rightharpoonup v) \quad ,$$

means for every $\ell \in X'$,

$$\lim_{n\to\infty} \ell(v_n) = \ell(v) \qquad . \tag{7.4}$$

Clearly $v_n \to v$ ensures $v_n \rightharpoonup v$, but not conversely.

DEFINITION 7.2. *A functional* A *on* X *is weakly continuous if whenever* $v = \text{weak} \lim v_n$, *then*

$$A(v) = \lim_{n\to\infty} A(v_n) \quad . \tag{7.5}$$

The functional is weakly lower semi-continuous if

$$A(v) \leq \lim_{n\to\infty} A(v_n) \quad . \tag{7.6}$$

The notion of weak lower semi-continuity arises when we look for local minima of a functional $A(v)$. Suppose that $\Omega \subset X$ is a ball on which A is bounded from below, namely

$$\inf_{v\in\Omega} A(v) = \bar{A} \quad .$$

Define a weakly convergent minimizing sequence $\{v_n\} \subset \Omega$ as a sequence such that $A(v_n)$ decreases monotonically to $\bar{A}$,

$$A(v_n) \searrow \bar{A} \quad , \tag{7.7a}$$

and such that there exists $v \in \Omega$ such that

$$v_n \rightharpoonup v \quad . \tag{7.7b}$$

If A is weakly lower semicontinuous, then the limit point v of a weakly convergent minimizing sequence satisfies

$$A(v) = \bar{A} \quad .$$

Hence if v is in the interior of Ω, v is a local minimum of A, see Figure 7.1. The question of when a weakly convergent minimizing sequence exists is a separate issue, dealt with in §8.

We remark that every functional ℓ in the dual X' is by definition weakly lower semicontinuous.

PROPOSITION 7.3. *Let* X *be a Hilbert space with norm* $\|\cdot\|$. *The map* $v \to \|v\|$ *is weakly lower semicontinuous.*

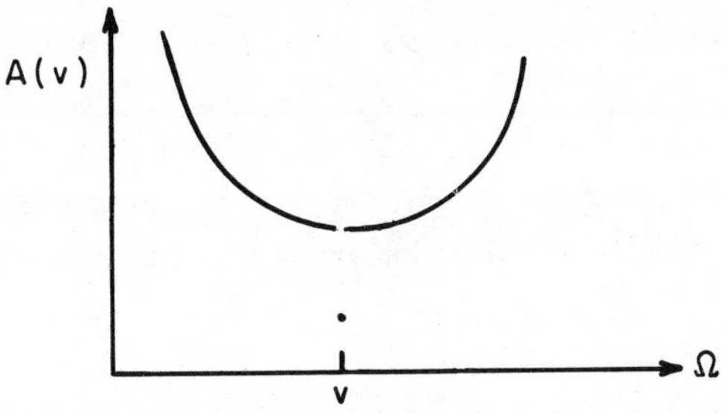

Figure 7.1. A lower semicontinuous function on Ω, near a local minimum v.

Proof. This follows from

$$\|v\|^2 = \lim \langle v_n, v\rangle + \lim \langle v - v_n, v\rangle$$

$$= \lim \langle v_n, v\rangle \leq \lim \|v_n\| \, \|v_n\| \quad . \tag{7.8}$$

If $v = 0$, then $\|v\| \leq \lim \|v_n\|$. If not, divide (7.7) by $\|v\|$.

The notion of extrema of a functional exists without defining a derivative. However, in order speak of the variational equations of A; and in order to identify the extrema with solutions to the variational equations, the derivative needs to be defined.

DEFINITION 7.4. *A functional* A *on a Banach space* X *has a directional derivative at a point* $v \in X$ *if for all* $h \in X$,

$$(D_h A)(v) = \lim_{\varepsilon \to 0} \frac{1}{\varepsilon} \left(A(v + \varepsilon h) - A(v)\right) \tag{7.9}$$

exists, and defines for fixed v, *a bounded linear functional on* X. *It is differentiable on an open set* $\Omega \subset X$ *if the derivative exists at all* $v \in \Omega$.

The relation between critical points (vanishing directional derivatives) and extrema (local minima or maxima) is given by the remark:

PROPOSITION 7.5. *Let the functional* A *on a Banach space* X *be differentiable in a neighborhood of an extrema* $\bar{v} \in X$. *Then for all* $h \in X$, $D_h A(\bar{v}) = 0$.

<u>Proof</u>. Consider for fixed h, $A(\bar{v} + \varepsilon h)$ as a function of ε, $|\varepsilon| < \delta$. Then $A(\bar{v} + \varepsilon h)$ has an extremum at $\varepsilon = 0$ and is differentiable there, so

$$0 = \frac{d}{d\varepsilon} A(\bar{v} + \varepsilon h)\Big|_{\varepsilon=0} = (D_h A)(\bar{v}) \quad . \tag{7.10}$$

As an element of X' for fixed v, $(D_h A)(v)$ is often denoted grad $A(v;\cdot)$.

If the functional is differentiable, then every extremum is a critical point, i.e., a solution to the variational equations grad $A(v;\cdot) = 0$. The uniqueness of extrema follows from strict convexity, a term we now define.

DEFINITION 7.6. *A functional* $A(\cdot)$ *on a Banach space* X *is convex if for all* $u \neq v \in X$ *and* $t \in (0,1)$

$$A(u + t(v-u)) \leq tA(v) + (1-t)A(u) \quad . \tag{7.11}$$

If the above inequality is strict, A *is said to be strictly convex.*

PROPOSITION 7.7. *Let* A *be a convex, once differentiable functional on a Banach space* X. *Then any critical point* $\bar{v}$ *of* A *is an absolute minimum:*

$$A(\bar{v}) = \inf_{v \in X} A(v) \quad . \tag{7.12}$$

If A *is strictly convex,* $\bar{v}$ *is unique.*

Proof. Let $\bar{v} \in X$ be a critical point of A. By (7.11), for any $u \in X$ and $t \in (0,1)$

$$A(\bar{v} + t(u-\bar{v})) - A(\bar{v}) \leq t(A(u) - A(\bar{v})) \quad . \tag{7.13}$$

Therefore, dividing both sides by t and letting $t \to 0$ we have

$$0 = \lim_{t\to 0} \frac{1}{t} (A(\bar{v} + t(u-v)) - A(v)) \leq A(u) - A(\bar{v}) \quad . \tag{7.14}$$

Hence $\bar{v}$ must be an absolute minimum of A on X. Suppose A is strictly convex and both $v_1 \neq v_2$ satisfy (7.12). Then

$$A(v_1 + t(v_2 - v_1)) < A(v_1) \quad , \tag{7.15}$$

which is a contradiction to $A(v)$ being the infimum.

The next proposition in this section relates convexity, differentiability and weak lower semi continuity.

PROPOSITION 7.8. *Let* A *be a convex functional with derivative on* X. *Then* A *is weakly lower semicontinuous.*

Proof. *Let* $\{v_n\} \in X$ converge weakly to $v \in X$. Using (7.11), we have

$$\frac{1}{t} (A(v + t(v_n - v)) - A(v)) \leq A(v_n) - A(v) \quad , \tag{7.16}$$

and hence

$$\text{grad } A(v, v_n - v) \leq A(v_n) - A(v) \quad . \tag{7.17}$$

Since $\text{grad } A(v,\cdot) \in X'$, it is weakly continuous. Therefore

$$0 \leq \lim_{n\to\infty} A(v_n) - A(v) \quad , \tag{7.18}$$

as claimed.

As an example, we prove that the function $v \to \|v\|^2$ is strictly convex.

PROPOSITION 7.9. *A positive quadratic functional* $v \to q(v,v)$ *on* X *is convex. If* q *is positive definite then it is strictly convex.*

Proof. The proof is by direct calculation,

$$\begin{aligned} q(tv+(1-t)u, tv+(1-t)u) &= t^2 q(v) + (1-t)^2 q(u) + t(1-t)(q(u,v)+q(v,u)), \\ &= tq(v) + (1-t)q(u) - t(1-t)q(u+v,u+v)), \\ &\leq tq(v) + (1-t)q(u) \quad . \end{aligned} \tag{7.19}$$

We end this section with the remark that a linear functional is convex.

VI.8. CALCULUS OF VARIATIONS

We establish sufficient conditions on a functional A defined on a reflexive Banach space X to achieve its minimum. A simple result is the following:

PROPOSITION 8.1. *Let* A *be a weakly lower semicontinuous functional defined on a closed ball* Ω *in a reflexive Banach space* X. *Then* A *achieves its minimum on* Ω.

We explain the undefined terms in the proposition: A set $\Omega \subset X$ is weakly closed if every weakly convergent sequence in Ω has its limit point in Ω. A set is weakly compact if every bounded sequence $\{v_n\}$ has a weakly convergent subsequence, converging to $v \in \Omega$. A Banach X space is reflexive if $X'' = X$.

PROPOSITION 8.2. *The unit ball of a Banach space* X *is compact if and only if* X *is reflexive. Every Hilbert space is reflexive.*

We do not prove Proposition 8.2, which can be found in any standard text in functional analysis. We remark further:

PROPOSITION 8.3. *The Banach spaces* L_p^k *are reflexive for* $1 < p < \infty$, *and* $0 \leq k < \infty$. *The dual of* L_p *is* $L_{p'}$, *where* $p' = p/(p-1)$, *and* $p < \infty$.

Proof of Proposition 8.1. Let $\{v_n\} \subset \Omega$ be a minimizing sequence for A, i.e., let $\{v_n\}$ satisfy (7.7a). Since Ω is bounded, $\{v_n\}$ is also bounded. Since X is reflexive, Ω is weakly compact. Hence $\{v_n\}$ has a weakly convergent subsequence $\{v_{n'}\} \rightharpoonup \bar{v} \in \Omega$. Hence $\{v_{n'}\}$ is a convergent minimizing sequence, and by weak lower semicontinuity

$$A(\bar{v}) = \bar{A} \quad . \tag{8.1}$$

Proposition 8.1 does not rule out the possibility that $\bar{v}$ may lie in the boundary of Ω. (For example, the exponential $x \to \exp x$ is C^∞ and strictly convex. Its minimum on every bounded interval $[a,b]$ is achieved at the endpoint a. There is no global minimum: in fact $\exp x \neq 0$, but $\inf \exp x = 0$.)

In order to assure that A has a minimum in X, stronger assumptions are necessary. One way to assure a local minimum exists for A is to prove that $A(v)$ grows as v approaches the boundary of Ω.

COROLLARY 8.4. *Let* A *be weakly lower semicontinuous functional defined on a reflexive Banach space* X . *If, on a ball* $\Omega = \{f: \|f - f_0\| \leq R\}$, *there exists* $v \in \Omega$ *such that*

$$A(v) < \inf_{\|f-f_0\|=R} A(f) \tag{8.2}$$

then A *achieves a local minimum in* Ω.

Proof. The condition (8.2) shows that the minimizing function $\bar{v}$ of the proposition is an interior point. Thus it is a local minimum.

We remark that (8.2) holds if A restricted to functions such that $\|f-f_0\| = R$, grows as $R \to \infty$. In that case A is said to be *coercive*.

PROPOSITION 8.5. *Let* A *be a weakly lower semicontinuous functional on a reflexive Banach space* X. *Let* $\gamma(t)$ *be a function such that* $\lim_{t\to\infty} \gamma(t) = \infty$. *Suppose there exist* b *such that*

$$A(v) \geq \gamma(\|v\|) - b \quad . \tag{8.3}$$

Then A *achieves its infimum on* X.

The picture to have in mind is that of Figure 8.1.

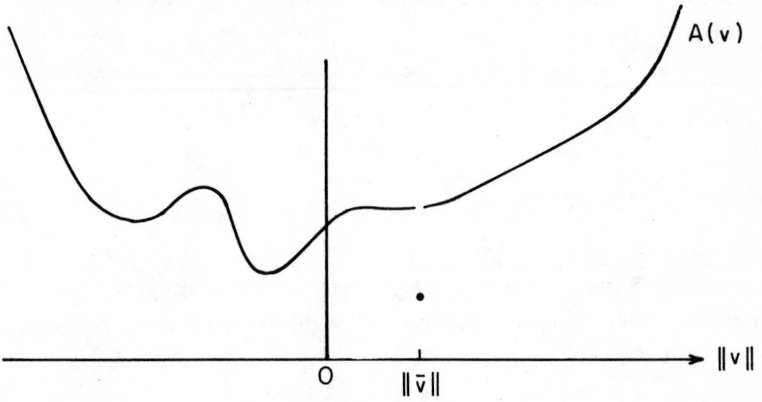

Figure 8.1. A(v) grows as $\|v\| \to \infty$. Weak lower continuity ensures a global minimum exists, Proposition 8.5.

We now give a sufficient condition for a minimum in terms of the derivative of A.

PROPOSITION 8.6. *Suppose that the functional* A *is weakly lower semicontinuous and strongly differentiable on* X. *If for some* $0 < R < \infty$ *and* $\delta > 0$

$$\inf_{\|v\|=R} (D_v A)(v) \geq \delta \quad , \tag{8.4}$$

then A *has a local minimum in the open ball of radius* R.

The picture to have in mind for Proposition 8.6 is that of Figure 8.2.

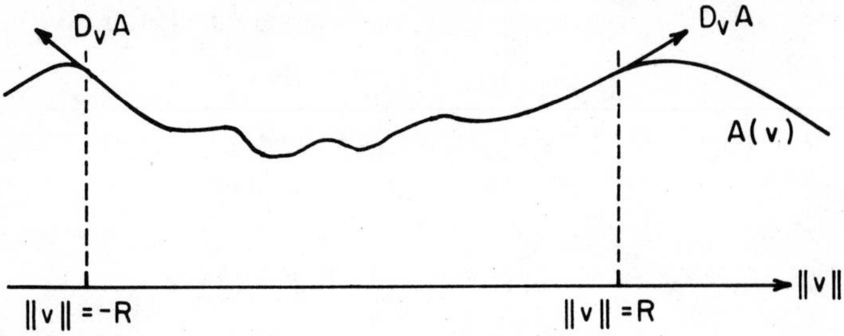

Figure 8.2. A satisfying hypotheses of Proposition 8.6.

Proof of Proposition 8.6. By Proposition 8.1 A achieves a minimum in Ω_R. It remains to be shown that the minimum is in the interior. Suppose the infimum $A(v)$ over $||v|| \leq R$ is achieved at some $\bar{v}$ satisfying $||\bar{v}|| = R$. Then for $t \in [-\varepsilon,\varepsilon]$, the function $A(\bar{v} - t\bar{v})$ is differentiable and

$$\frac{d}{dt} A(\bar{v} - t\bar{v})\Big|_{t=0} = -(D_{\bar{v}}A)(\bar{v}) \leq -\delta \quad . \tag{8.5}$$

Therefore, there exists $0 < \mu < \varepsilon$ such that

$$A(\bar{v} - \mu\bar{v}) < A(\bar{v}) \tag{8.6}$$

which is a contradiction since

$$||\bar{v} - \mu\bar{v}|| = (1 - \mu)R < R \quad .$$

VI.9. LOCAL REGULARITY

The purpose of this section is to present some standard regularity theorems for elliptic PDE's. We refer the reader to [Morrey] and the references therein for greater detail and for the proofs of the following statements.

Let $V \subset \mathbb{R}^d$ be a bounded, open set with smooth boundary $\partial\bar{V}$. We are concerned with the regularity of a function $u \in H_1$ which satisfies

$$0 = \int_V [\nabla_j \zeta (a^{jk}\nabla_k u + b^j u + e^j) + \zeta(c^k \nabla_k u + gu + f)] \quad \text{for all} \quad \zeta \in C_0^\infty(V). \tag{9.1}$$

We shall assume that the a^{jk} are bounded and satisfy

$$m|\lambda|^2 \le a^{jk}\lambda_j\lambda_k, \quad |a(x)| \le M \qquad \text{for a.e.} \quad x \in V \text{ and all } \lambda. \tag{9.2}$$

Conditions on the coefficients b,c,g,e,f will be added as needed.

THEOREM 9.1. *Suppose the following statements are satisfied:* $a^{jk} \in C^1(B)$, $b^j \in L^1_{d/(d+2)}(B)$, $e^j \in H_1(B)$ *is bounded,* $g \in L_d(B)$, $f \in L_2(B)$, *and* $\int_{B_r(x_0)\cap B} (|\nabla b|^2 + |g|^2)^{d/2} \le z_0 r^{\mu_1}$, $\mu_1 > 0$ *for each open ball* $B \subset V$, *point* $x_0 \in B$ *and* $r > 0$, *and* z_0 *is a constant.*

Then $u \in H_2(B)$ *for each such* B *where it satisfies (almost everywhere) the equation*

$$-\nabla_j(a^{jk}\nabla_k u + b^j u + e^j) + c^j \nabla_j u + gu + f = 0 \quad . \tag{9.3}$$

Further, if e *and* f *satisfy the above conditions on* V *and* $u \in L'_{2(0)}(V)$, *then* $u \in H_2(V)$. *If the coefficients are* C^∞, *then* $u \in C^\infty$.

Theorem 9.1 is the basic tool for proving the regularity of a solution to a nonlinear PDE. The usual technique is to use Theorem 9.1 in an induction proof on the largest integer k such that $u \in H_k(B)$ for all open balls $B \subset V$.

An *a priori* upper bound on a subsolution to (9.1) is given in

THEOREM 9.2. *Suppose that the following statements are satisfied:*

(1) $u \in H_1(B)$ *for each ball* $B \subset V$.

(2) *For each* $\zeta \geq 0$ *in* $C_0^\infty(V)$,

$$\int_V \left\{ \nabla_j \zeta (a^{jk} \nabla_k u + b^j u) + \zeta (c^j \nabla_j u + gu) \right\} \leq 0. \tag{9.4}$$

(3) *For* $x_0 \in V$ *and* $B_r(x_0) \subset V$, *and some constant* z_0,

$$\int_{B_r(x_0)} (|b|^2 + |c|^2 + |g|^2)^{d/2} \leq z_0 r^{\mu_1} \quad , \quad \mu_1 > 0.$$

Then u *is bounded in each domain* $B \subset V$ *and*

$$|u(x)|^2 \leq z_1 a^{-d} \int_{B_{R+a}(x_0)} |u(y)|^2 \quad , \qquad x \in B_{R+a}(x_0)$$

$$0 < a \leq R \quad , \qquad B_{R+a}(x_0) \subset V \quad . \tag{9.5}$$

REFERENCES

CHAPTER I

1. A.A. Abrikosov, *JETP* 5, 1174 (1975).

2. N. Ashcroft and D. Mermin, *Solid State Physics*, Holt, Rinehard and Winston (1976).

3. M.F. Atiyah and R. Bott, to appear.

4. M.F. Atiyah, V.A. Drinfeld, N.J. Hitchin and Yu.I. Manin, *Phys. Lett.* 65A, 185 (1978).

5. M.F. Atiyah, N. Hitchin and I. Singer, *Proc. Nat. Acad. Sci.* (U.S.A.) 74, 2662 (1977). *Proc. Roy. Soc.* (London) A362, 425 (1978).

6. M.F. Atiyah and J.D.S. Jones, *Commun. Math. Phys.* 61, 97 (1978).

7. M.F. Atiyah and R. Ward, *Commun. Math. Phys.* 55, 117 (1977).

8. A.A. Belavin, A.M. Polyakov, A. Schwartz and Y. Tyupkin, *Phys. Lett.* 59B, 85 (1975).

9. J.P. Bourguinon and H.B. Lawson, Jr., *Commun. Math. Phys.*, to appear. Proc. Nat. Acad. Sci. (USA) 76, 1550 (1979).

10. D. Brydges and P. Federbush, *Commun. Math. Phys.* 73, 197 (1980).

11. D. Brydges, J. Fröhlich and E. Seiler, *Commun. Math. Phys.* 71, 159 (1980), *Ann. Phys.* 121, 227 (1979), and to appear.

12. C. Callan, R. Dashen and D. Gross, *Phys. Lett.* 66B, 375 (1977).

13. N. Christ, E. Weinberg and N. Stanton, *Phys. Rev.* D18, 2013 (1978).

14. S. Coleman, Lectures at the 1975 Erice Summer School, Plenum (1977).

15. E.F. Corrigan, D.B. Fairlie, R.G. Yates and P. Goddard, *Commun. Math. Phys.* 58, 223 (1978).

16. P.A.M. Dirac, *Proc. Roy. Soc.* A133, 60 (1931).

17. V.A. Drinfeld and Yu.I. Manin, *Commun. Math. Phys.* 63, 177 (1978).

18. F. Dunlop and C. Newman, *Commun. Math. Phys.* 44, 223 (1975).

19. J. Ellis, Lectures at Schladming, Austria (1979).

20. R. Flume, *Phys. Lett.* 76B, 593 (1978).

21. H. Georgi and S. Glashow, *Phys. Rev. Lett.* 32, 438 (1974).

22. V.L. Ginzburg and L.D. Landau, *Zh. Eksp. Theor. Fiz.* 20, 1064 (1950).

23. J. Glimm and A. Jaffe, *Quantum Physics*, to appear (1981).

24. J. Glimm, A. Jaffe and T. Spencer, *Ann. Phys.* 101, 610, 631 (1976).

25. P. Goddard, J. Nuyts, D. Olive, *Nuclear Phys.* B125, 1 (1977).

26. A. Guth and S.-H. Tye, *Phys. Rev. Lett.* 44, 631, 963 (1980).

27. G. 't Hooft, *Nucl. Phys.* B79, 276 (1976), and unpublished.

28. R.P. Huebener, *Magnetic Flux Structures in Superconductors*, Springer (1979).

29. L. Jacobs and C. Rebbi, *Phys. Rev.* B19, 448b (1979).

30. A. Jaffe in *Proceedings Helsinki ICM* (1978).

31. S. Mandelstam, *Phys. Rev.* D19, 2391 (1979).

32. D. Olive, in *Proceedings Lausanne 1979*. Springer LNP116 (1980).

33. R.D. Parks (editor) *Superconductivity* (2 Volumes), Marcel Dekker (1969).

34. B. Plohr, Thesis. Princeton (1980).

35. A.M. Polyakov, *JETP Lett.* 20, 194 (1974). *Nucl. Phys.* B120,429 (1977).

36. M.K. Prasad and C.M. Sommerfield, *Phys. Rev. Lett.* 35, 760 (1975).

37. J. Preskill, *Phys. Rev. Lett.* 43, 1365 (1979).

38. D. Saint James, G. Sarma and E.J. Thomas, *Type II Superconductivity*, Pergamon (1969).

39. A. Schwartz, *Phys. Lett.* 69B, 172 (1977).

40. J.R. Schrieffer, *Theory of Superconductivity*, Benjamin (1964).

41. C. Taubes, *Commun. Math. Phys.* 75 (1980).

42. E. Witten, *Phys. Rev. Lett.* 38, 121 (1977).

43. T.T. Wu and C.N. Yang, *Phys. Rev.* D12, 3845 (1975).

44. C. N. Yang and R.L. Mills, *Phys. Rev.* 96,191 (1954).

CHAPTER II

1. E. Cartan, *Ouvres complètes*, Part I, Vol. 2, 1081. Gautier Villars (1952).

2. S. Coleman, see I.14.

3. G. Derrick, *J. Math. Phys.* 5, 1252 (1964).

4. F. Englert and P. Windey, *Phys. Rev.* D14, 2728 (1976).

5. P. Forgács and N. Manton, *Commun. Math. Phys.* 72, 15 (1980).

6. R.T. Glassey and W. Strauss, *Commun. Math. Phys.* 67, 51 (1979).

7. P. Goddard, J. Nuyts and D. Olive. See I.25.

8. P. Goddard and D. Olive, Phys. Reports 41, 1357 (1978).

9. S. Helgason, *Differential Geometry and Symmetric Spaces*, Academic Press (1962).

10. T. Johnsson, Thesis, Harvard (1980).

11. S. Kobayshi and K. Nomizu, *Foundations of Differential Geometry*, Interscience (1963).

12. J. Madore, *Commun. Math. Phys.* 56, 297 (1977).

13. M. Mayer, Lectures Mexico City (1979), to appear.

14. M. Monastyrskii and A.M. Peremelov, *JETP Lett.* 21, 43 (1975).

15. L. Pontryagin, *Topological Groups*, Princeton (1946).

16. R. Schlafly, "A Chern Number for Gauge Fields on $\mathbb{R}^4$," to appear.

17. R. Schoen and S.-T. Yau, *Ann. Math.* 110, 127 (1979).

18. E.H. Spanier, *Algebraic Topology*, McGraw-Hill (1966).

19. R. Stora, Lectures on Gauge Theories.

20. Yu. S. Tyupkin, V.A. Fateev and A.S. Schwartz, *Teor. Mat. Fiz.* 26, 397 (1976).

21. K. Uhlenbeck, "Removable singularities in Yang-Mills fields," to appear. "Connections with L_p bounds on curvature," to appear

22. E. Witten, see I.42.

CHAPTER III

1. E.B. Bogomol'nyi, *Sov. J. Nucl. Phys.* 24, 449 (1976).
2. F. Dunlop and C. Newman, see I.18.
3. P. Griffiths and J. Harris, *Principles of Algebraic Geometry*, Wiley (1978).
4. C. Morrey, *Multiple Integrals in the Calculus of Variations*, Springer (1966).
5. C. Taubes, *Commun. Math. Phys.* 72, 277 (1980).
6. C. Taubes, *Commun. Math. Phys.* 75 (1980).

CHAPTER IV

1. S. Adler, *Phys. Rev.* D17, 3212 (1978).
2. S. Adler, *Phys. Rev.* D19, 1168 (1978).
3. F.A. Bais, *Phys. Rev.* D18, 1206 (1978).
4. E.B. Bogomol'nyi, *Soviet J. Nucl. Phys.* 24, 449 (1976).
5. P. Forgács and N. Manton, see II.5.
6. G. 't Hooft, see I.27.
7. P. Houston and L.O'Raifeartaigh, to appear.
8. S. Kobayshi and K. Nomizu, *Foundations of Differential Geometry*, Interscience (1963).
9. A. Leznov and M. Saveliev, *Lett. Math. Phys.* 3, 207 (1979).
10. A. Leznov and M. Saveliev, *Commun. Math. Phys.* 74, 111 (1980).
11. N.G. Lloyd, *Degree Theory*, Cambridge Press (1978).
12. C. Morrey, see III.4.
13. A.M. Polyakov, see I.35.
14. M.K. Prasad and C.M. Sommerfield, see I.36.
15. L.O'Raifeartaigh, *Nuovo Cim. Lett.* 18, 205 (1976).
16. C. Rebbi and P. Rossi, to appear.
17. D.M. Scott, to appear.

18. N. Steenrod, *Topology of Fibre Bundles*, Princeton (1951).

19. C.H. Taubes, "Existence of multimonopole solutions to the static, SU(2) Yang-Mills-Higgs equations in the Prasad-Sommerfield Limit," to appear. "Existence of multimonopole solutions for Arbitrary Simple Gauge Groups," to appear.

20. C.H. Taubes, Thesis, Harvard (1980).

21. R. Weder, to appear.

22. E. Weinberg, *Phys. Rev.* D20, 936 (1979).

23. E. Weinberg, to appear.

24. D. Wilkinson and A. Goldhaber, *Phys. Rev.* D16, 1221 (1977).

CHAPTER V

1. P. Mitter, Lectures Cargèse, Plenum, to appear.

2. C. Morrey, see III.4.

3. R. Palais, *Foundations of Global Nonlinear Analysis*, Benjamin (1968).

4. N. Steenrod, see IV.18.

5. C. Taubes, see IV.20.

6. K. Uhlenbeck, see II.21.

CHAPTER VI

1. R.A. Adams, *Sobolev Spaces*, Academic Press (1975).

2. M. Berger, *Nonlinearity and Functional Analysis*, Academic Press (1977).

3. R. Courant and D. Hilbert, *Methods of Mathematical Physics*, Vol. II, Interscience (1962).

4. D. Gilbarg and N.S. Trudinger, *Elliptic Partial Differential Equations of Second Order*, Springer (1977).

5. E. Hopf, Sitz, Ber. Preuss. Akad. Wiss. Berlin, Math. Phys. Kl. 19, 147 (1927).

6. S. Kobayshi and K. Nomizu, see II.11.

7. C. Morrey, see III.4.

8. R. Palais, see V.3.

9. M. Reed and B. Simon, *Functional Analysis*, Academic Press (1972).

10. E.M. Stein and G. Weiss, *Introduction to Fourier Analysis on Euclidean Spaces*, Princeton (1971).

11. M.M. Vainberg, *Variational Method and Method of Monotone Operators in the Theory of Nonlinear Equations*, Wiley (1973).

This work was supported in part by the National Science Foundation under Grant PHY 79-16812.